Sensi/able Spaces

Sensi/able Spaces
Space, Art and the Environment Proceedings of the SPARTEN conference, Reykjavík, June 1st and 2nd, 2006

Edited by

Edward H. Huijbens and Ólafur Páll Jónsson

CAMBRIDGE SCHOLARS PUBLISHING

Sensi/able Spaces: Space, Art and the Environment Proceedings of the SPARTEN conference, Reykjavík, June 1st and 2nd, 2006, edited by Edward H. Huijbens and Ólafur Páll Jónsson

This book first published 2007 by

Cambridge Scholars Publishing

15 Angerton Gardens, Newcastle, NE5 2JA, UK

British Library Cataloguing in Publication Data
A catalogue record for this book is available from the British Library

ISBN 1-84718-323-9; ISBN 13: 9781847183231

TABLE OF CONTENTS

PREFACE

I

On the 1st and 2nd June 2006 over 100 people took part in a conference and art exhibit under the heading Sens^{i}/$_{a}$ble Spaces: Space, Art and the Environment. The conference and art exhibit were held at the University of Iceland, whilst related events were took place at various locations around the city of Reykjavík.

The event's programme started in the afternoon of 31st May with a boat trip to the small Viðey island just off Reykjavík. On the island three Icelandic artists; Kristinn Hrafnsson, Rósa S. Jónsdóttir and Hannes Lárusson, guided the group around the artwork of Richard Serra and Site-ation International on the island. After the walk an evening reception in Viðeyjarstofa, the island's main building, was hosted by Reykjavík city council and Visit Reykjavík. On June 1st an opening panel was held at the Reykjavík Art Museum. From there, people moved to the University of Iceland where the minister for the Environment, Mrs. Sigríður Anna Þórðardóttir, opened an art exhibit along with presentation sessions. In the evening a reception in the Main Building of the University of Iceland was hosted by the rector of the University. On June 2nd sessions carried on for the day, but as they ended the formal programme ended whilst the art exhibit stayed open for the rest of the summer.

There were four main organisors for this event. Dr. Edward H. Huijbens, then working at the University of Iceland was the main coordinator and project manager whilst Paul Hackett, a PhD student at Gloucestershire University at Cheltenham, took care of curating the art exhibit. Dr. Ólafur Páll Jónsson, a lecturer at the Iceland University of Education, and Dr. Karl Benediktsson, a reader at the geography department of the University of Iceland, were pivotal in executing the tasks set out for the fruition of the event.

Along with the hard work of all the four mentioned above the event was made possible through the support of the Ministry of Education, Science and Culture along with the Ministry of the Environment. The Reykjavík Arts Festival, with Reykjavík City Council and the Reykjavík Art Museum supported the event generously as did the University of

Iceland and the Faculty of Economics and Business Administration. Finally the British Council lent support to make the event possible. All our supporters are hereby thanked for their invaluable help and good will.

II

The title "Sens$^{i}/_{a}$ble Spaces" was meant to suggest the way in which space, art and the environment interlace and interact, dealing with the perception and conception of spaces in the built as well as natural environment. The aim of the event was to bring together a wide range of academics, from the physical sciences, social sciences and humanities, as well as artists and activists who have an interest in the way spaces is sensed, understood and reconfigured.

The rational presented for the event was that spaces today are continually being reconstituted and reformulated in various ways, often relying on notions of what is sens*i*ble, narrowly defined by groups with an ideological agenda of some kind or vested economic interests. What we argued is that these sens*i*ble factors often obscure and ignore notions of the sens*a*ble—that which people perceive through the senses while being-in spaces. These sens*a*ble spaces, in reference to the built or natural environment or all those habitually lived spaces, are premised upon phenomenology—the study of how things appear before the conscious mind.

Space as a subject of academic interest has a somewhat curious status. On the one hand, it is the subject of higly technical and abstract theories, rooted in approaches as distinct as the metaphysics of Aristotle, the phenomenology of Martin Heidegger and Edmund Husserl to analytic approaches in mereology and topology. On the other hand, space is among the most mundane things one can imagine. Everybody has an experience of space, and most people know quite a lot about how to create spaces or change their character, epitomised, e.g. in popular programmes on interior design, theatre art and the popular art of *feng-shui*. The task of the academics is, then, both to fit the concept of space into a wider theoretical framework, linking it to notions such as *movement*, *change*, *consciousness*, *environment*, *place* and, as this present volume emphasizes, the sens*i*ble and the sens*a*ble. But while doing this, the academic must also be faithful to the every-day experience of people engaging with spaces. Due to this fact, space is a topic equally of various academic fields, such as geography, philosophy, sociology, anthropology, physics, bilology, and many more. But space is also the subject of—or a frame for—any artists, whose work is neither academic, in any standard sense of that term, and yet heavily theoretical or speculative.

III

The volume is divided in three parts “Space”, “Art” and “Environment”. Leading off into each part are the keynote addresses presented in the Reykjavík Art Museum at the opening event of the 2006 conference and art exhibit. The first part, “Space”, is headed by a paper by Doreen Massey, a professor of geography at the Open University in England. The second part, “Art”, is headed by a paper by Andrew A. Stonyer, professor of sculpture at the University of Gloucestershire at Cheltenham. The third part, “Environment”, is headed by a paper by Gísli Pálsson, professor of anthropology at the University of Iceland.

In the first part are five papers dealing with differnt aspects of space. Anna Lísa Runólfsdóttir argues that both form and content are significant in the experience of a space. Drawing on Bourdieu’s *habitus* she argues that our immediate environment conditions our behaviour and contributes to the maintenance of cultural values. Objects and spaces are embodied with meaning, which is transmitted in a non-discursive way, through practice. The ideology thus manifested within space and material culture supports values and social structure. What she poses as a question is how *habitus* is affected when significant technical advances are introduced to the home, with reference to the social structure embedded in the Icleandic turf houses of the past?

In order to discuss spaces that offer alternative meanings and experiences, Julia Neváres draws on the notion of *agoraphobia* from urban studies, the notion of *felicitous spaces* from Bachelard, the notion of *heterotopias* from Foucault and the notion of *rhythm analysis* from Lefebvre. Daydreams and spaces identified with our solitude, the spaces that offer momentary glances into surprise and unexpectedness, the urban uncanny are some of the ideas that guide the selection of urban spaces in her paper. To understand spaces that offer the possibility for the development of the imagination is what people perceive to be spaces where they can be engaged in ways that provide comfort or refuge, spaces that encourage exploration and even dissent. These spaces can be reconstituted and reformulated through visual perception and movement, in the connectedness derived from journeys. Accordingly, the spatial trajectories that frame the meanings of visually reading a landscape is coherent with a notion of relatedness of different spaces that signify and provides meaning to urban space.

Julienne Hanson, Reem Zako, Helen Bendon and Jessica Thom talk about housing and neighbourhood developments. They begin by outlining how the late twentieth century was shaped by the belief that it is important to create communal open spaces that can provide a physical focus for local communities. This has given rise to a range of modern housing typologies that are based on an ethos of building community through design, using concepts such as a *privacy gradient*, the separation of pedestrian and vehicular movement and the notion of *enclosure*, amongst others, to shape the public open space of the estate. Their paper examines the contribution of such public spaces to community awareness in twenty-eight English city-centre housing estates with differing housing morphologies, built between the middle of the nineteenth century and the early twenty-first century. As a manifestation of the complex social relationships that take place within the domain of public space, they suggest that the above concepts and their physical realisation need to be complemented by careful consideration of social notions such as the constitutedness of the spaces, co-presence and surveillance.

Elisebha Fabienne Platzer focuses on speech as the topical semiotic view on spaces. She compares the semiotic transformation of the city in artworks in the landscape through a three part approach. First through looking at the city as a 'textum' of signs in the acceptation of texture and testimony. Secondly through looking at the city as a network of actions, interactions, gestures and thoughts and thirdly through art, aiming to point out by the conscious and unconscious practices that trace signs, eliminate striations, to create an "espace lisse" in order to rewrite space by inventing 'Verfremdungseffekte', deautomatized language.

Thomas Heyd presents an invisible network of human-nature relations punctuated by native boulder structures at its nodes on the Northern Plains of North America. The common name of these structures presently is 'medicine wheels'. Despite considerable archaeological research, their original meaning is difficult to ascertain. They certainly have an important significance for contemporary First Nations people. In his paper he explores the significance of these landmarks for non-Native peoples who have appropriated the area since colonization. He argues that there are good reasons to consider these structures as artworks and, furthermore, that these structures have the function of *heterotopias*, that is, sites that question and interrogate all other places.

The section on art contains five chapters in addition to the introductory chapter by Andrew A. Stonyer. Manuela Silvia Campanini and Hrafnhildur Schram, both working in the field of environmental sustainability, present a study of people's colour perception of the

Icelandic lansdscape. They evaluated the perceived colours of a group of 270 Italian tourists in Iceland through a survey and compared them to standardised colour pallettes. These tourists travelled in Iceland with organized group trips, 6 or 8 days in length, in the period between June 3rd and August 30th, 2005.

Antoine Capet aims to 're-place' the war time photographic reportage of Bill Brandt in a thinly-disguised plea in favour of Reconstruction on new bases after the war. This 're-placing' he aims for in the general context of Bill Brandt's perception of British 'space', the *Homes for Heroes* series coming after the pre-war reportage in the mining districts of the North and immediately before his post-war 'Literary Britain' landscapes, all of which he documents in his paper.

Michael Shiell gives an example of the devolution of an ephemeral environmental artwork in western New South Wales. Begun in 2001, the work entitled 'Bone Circle/Bone Fire' has considerably devolved. He argues that with this devolution however there has been a development in the significance of the space for the small community that uses the area. He builds his paper on the argument that unlike permanent, solid works designed to withstand the rigours of time, ephemeral environmental works exist within time's cycle. They are not intended to be a massive imposition within a space but rather an element of that space that will pass. This brevity of life when coupled with a lack of any formal de-installation process means that the works departure is more like the gentle passing of a life. Unlike the finality of removing work from a site, devolution allows the work to evolve into something else. In this case the intent of the artist and thus the original work may be lost but the constituent elements of the work can remain. This also allows for the continued evolution of society's connection to that space through that work.

Shauna Laurel Jones talks about cinema and the portraying of spaces, specifically the film Nói Albinói. She argues that by means of the unconventional and uncomfortable space Pétursson, the director, has created, Nói Albinói calls attention to the filmic medium and to the act of looking itself, as well as to the act of viewing and experiencing space. Given the deleterious consequences of the sense of disconnect many Westerners have from the natural environment, it is critical that we begin to investigate how contemporary visual culture addresses this issue—and how it might be used to reshape our understanding of landscape to better promote our reconnection with it.

Serge Tampalini talks through the notion of 'affective space' and argues that a possible first step in conceptualizing affective space would be to try and detach what one sees from what one knows. A second step would be to inscribe space in such a way as to give it the potential of 'looking back'. He draws his examples from theatre design, which he argues is already inscribed; doors, tables, chairs, stage properties etc. are already a kind of inscription/label; we do not see a chair, without relating it to the idea of a chair. This type of inscription functions as part of what he calls 'effective space', and is exploited by the practice of design as decoration. In order to have space 'look back' he argues we need to rethink the idea of inscription. Unlike the inscriptions of effective space, the inscriptions of affective space are twice removed from their counterparts in reality. Their concerns are more than meaning and function – affective space has what might be called an excessive condition; one that participates in the conventions of space and time.

Moving from the theatre to the environment as a whole, Gísli Pálsson leads off into the last part of the edited volume. There Pierre Auboiron talks about the use of light as environmental tracer. In his paper he explains how artists and architects intend to make buildings fit back into their historical and socio-geographical environment and provide the public with a more accurate vision of the city and its main districts and buildings. He tells of how they have based their use of light on a social or environmental indexation in order to make concepts and features of everyday life visible and more tangible. Thereby they try and fight against the reduction of interpersonal communication in today's urban life which is the result of an increasing visuality. He argues that these artists and architects materialise phenomena we cannot perceive anymore because we have developed our visual sense to the detriment of our other senses. This could be compared to the use of tracers in biology which reveal invisible processes or organisms.

Giovanni Tuzet talks about the distinction made between culture and nature and questions this distinction through questioning the order in which they are presented. A distinction Maria Tamboukou questions in terms of urban and rural spaces. In her paper she looks more specifically into gendered perceptions of urban and rural spaces as inscribed in the epistolary narratives and paintings of Gwen John, a British artist who studied at the Slade School of Art and lived and worked in the UK and in France in the first half of the twentieth century. She argues that if we are always outside ourselves and open to the world, then narrative articulations of the spatial constitution of the female self as it surpasses the inner/outer division, can offer insights into gendered configurations of sens^{i}/$_{a}$ble spaces.

Kenichi Nonaka and Nozomi Yanahara present a study aimed at grasping the relationship and practices that are formed through interactions and links between creatures, humans and the geographical landscape and consider how to depict it. They attempt to develop means of representing it in the form of illustration to make it understood, drawing on the Japaneese manga cartoon tradition.

Paola Jíron presents her take on time-geography and the way it can analise the everyday spatiality of people in Santiago in Chile. She wants to understand how people signify daily encountered places through adopting innovative methodologies that will not only provide a verbal description of these places, but also a visual and spatial expression to understand their mobility as well as their meaning to urban practices. This paper proposes actualising time space mapping by complementing them with everyday diaries, interviews and photography to have a more comprehensive way of understanding what these spaces are, how they are created, and more importantly how they expand or restrict people experience of the city.

Christain Denker asks why we should reflect on free nature in order to understand our behavior towards our humanized environment, if the reflection on aesthetic and ethic values focalizes on art and culture? He proposes to answer this question by outlining that the idea of free nature has an emotional impact that is a crucial addition to rational reflection on the impact of human activity and perception on the environment.

We hope that the papers presented in this volume demonstrate differing takes on space, art and the environment inspiring readers to allow space's presence to be felt in and on their bodies. The project detailing the role of space, art and the environment in our lives is now just starting. Next we invite scholars and artists from a range of relevant disciplines to engage with *Vital rhythms – arctic timespace*. This Academic Art Conference is to be held at the University of Akureyri, in Iceland, beginning June 2008. There we encourage artists to team up with an academic to produce a site-specific work prior to and during the conference to be displayed on site. That event draws on the success of *Sensi/able Spaces*, but also a conference hosted by the Reykjavík Academy *Images of the North* held in Reykjavík in combination with The Reykjavík Winter Art Festivals, earlier in 2006.

The aim of *Vital rhythms* is to bring together key actors and research institutes in order to constitute networks throughout the northern periphery furthering the thoughts and issues presented in this volume. These networks are based on an understanding and sensibility to the delicate environments of the North and are to promote discussion on how these can be used in a sustainable manner. The focus of this event is on timespace and the way in which it is expressed through rhythms and the seasonality of everyday life in the North, in order to articulate the role of the material, especially art objects and artefacts, in people's everyday lives and how these are integral to the fabric of timespace in the north.

We would hereby like to thank all the contributors for their hard work both in presenting at the conference and writing up their presentations in chapter form for this publication. On the last pages of this volume notes on each of the contributors in terms of their education and recent work are presented. We would as well like to thank the anonymous referees that reviewed the articles prior to publication.

Edward H. Huijbens
Ólafur Páll Jónsson

Akureyri and Reykjavík

1st June 2007

PART I

SPACE

CHAPTER ONE

MOVING SPACES

DOREEN MASSEY

Space is alive. It is an ongoing construction of our interrelatedness, human and nonhuman. It is the dimension of multiplicity, the dimension of a simultaneity of innumerable stories-so-far. And places are the locations of the meeting-up of those stories (or of particular sets of them), articulations of a multiplicity of trajectories. 'Stories' and 'trajectories' speak of energy, of flow. All is movement. 'Multiplicity' requires that we negotiate our interrelatedness, for those intersecting trajectories may clash, contradict each other, nestle together in companiable fashion, or require the long hard effort of establishing mutual comprehension.

Given this, space and place are not terms, nor realities, that should be counterposed. Yet so often they are. 'Place' is so often evoked as 'real', 'authentic', 'everyday', 'human-scale', 'meaningful' … as against space which is in contrast figured as an abstract dimension. In this imagination, space is meaning*less*, *in*authentic, devoid of resonance and somehow beyond the scale of the human. This may be, perhaps, a peculiarly modern, or modernist, understanding of space but it is powerful. And both it and the counterposition with place that necessarily accompanies it are problematical.

In a conference, exhibition, and discussion devoted to sensible/sensable space it is important from the outset to recognise that 'space', even space simply as a dimension (that is, rather than particular spaces), is just as humanly produced as is place. The everyday is global now, and the global is everywhere local. On the one hand global space is produced in places; on the other hand places themselves are articulations of, meetings of, spatially wider relations. No particular place (Iceland maybe), nor meditations upon it, can provide a safe haven from the wider world.

It is particularly important to insist on these arguments in an age of globalisation, and in an age when the political Right so often seizes upon the specificities of place to argue for exclusivities (the barriers, the walls, the rhetorics of nationalism) as part of its political cosmology. However it is also important because there can be, on the political Left too, a certain romanticisation of the local place. In an era such as the present that cannot be afforded. Rather it is important to develop and work within a different kind of political cosmology, in which local places are specific, yes (and this remains supremely important, both presenting us, concretely, with the fact of difference and offering moments of exquisite pleasure) but where such specificity is the product of openness as much as of locally-placed entanglement. This would also be a political cosmology in which a different kind of globalisation could be imagined, in which, also, that wider space, even space simply as dimension, could itself have meaning and import. This is, perhaps, an element of what sensible/sensable space is reaching towards.

*

A walk across Viðey in the wind, with a host of people one had still to begin to encounter (a real meeting-up of trajectories) was provocative of many such meditations. And the encounter with artworks added to the complexity of the production of this particular here-and-now, generating new thoughts/trajectories in its turn. One such encounter took place in a wooden structure, light, airy and open to the sea.[1] Screwed-up bits of paper littered the floor. We (the group) and the artwork were introduced to each other. The work was a telling evocation of contrast between places, between the lack of water in Spain (whence the artist hails, and where lemon trees are dying in the drought) and the apparent abundance of water in Iceland. ‘This is not a dry world’ it was entitled (setting off the reflection that wars over water may well come to be fought that are as violent as the current contests over oil). The dessication of the screwed-up paper was powerful against the endless sea beyond. Yet as in so many encounters, the products are unpredictable, they wander off in new directions. Before our introduction (before knowing about the artist’s intentions, before knowing about the naming, before knowing about the dying of the lemon trees), before words had come between me and it, the materiality of the screwed-up balls of paper scattered before a cold, restless sea, generated (in me) something quite different. Again another place was invoked, this time another northern place – Spitsbergen, the main island of the archipelago of Svalbard. And the paper was litter, detritus. If you go ashore in Svalbard you take with you large hessian bags, and in this place so far from ‘civilisation’, a place where many go to

---

[1] This was the work of Tony Ceballos.

get away, you walk the shoreline. And you gather plastic rings, and bottle tops, bits of junk and rubbish. Litter, which has arrived here on the vast flow of the North Atlantic Drift, the rubbish carelessly strewn from cruise-boats off Florida maybe, or in the Caribbean. The bellies of the fulmars and the polar bears are filling up with plastic.

In this new train of thought water is flow, movement. It is one of the (myriad) means of interconnectedness, human and beyond human, around the planet. The contrasts between places, so beautifully evoked between Spain and Iceland, are also the generators of exchange, interrelation, and mutual production. On that wild late afternoon, on an island very few of us had ever visited before, in a moment of new encounters, thoughts were provoked not simply of benign interconnectedness but also of the more demanding proposition of the responsibilities of place beyond place.

*

In the artworks more generally, there seemed to be a persistent theme, a weighty undertow, that wove itself around, on the one hand, the rural and the (so-called) natural and, on the other hand, memory and the past. There was play on the impossible-to-disentangle interworkings of natural and humanly-produced. There were recallings of Iceland's (humanly-produced) traditions—the cairn, the knitted sweaters Gísli writes about in this collection.[2] Much of it raised questions about the positioning of place in general and of Iceland in particular.

It is, usually, difficult to see anything through the mist and drizzle as the aeroplane sinks down towards Keflavík. But emergence from the airport presents the newly-arrived traveller with much to ponder. Most spectacularly, there is a nature that is as new, and mobile, as anywhere on the planet. Meanwhile on the human front what most confronts is the brute presence of a US airbase and an aluminium plant.[3]

One of the strongest images of Iceland is indeed the newness and mobility of the natural. For an Englishwoman the effect is startling. No recourse is possible here, as is so frequently attempted at home, to 'nature' as stable eternity; nature as foundation. We were standing among the towering basalt pillars of Áfangar on Viðey when someone asked, in relation to its complex calculations, 'was all this done in relation to high tide or to low?'[4] In other words, in the midst of all of this, what is zero?

---

[2] References to the artworks in Viðey, see introduction.
[3] The airbase is now, of course, evacuated.
[4] Áfangar (Stages) is by Richard Serra.

The question can in fact be taken further for not only, as we know, is the sea pulled, to and fro, by the moon, but so too is the land. The middles of continental landmasses rise and fall in response to the same forces and by considerable distances each day. There is, in other words, no stable point. There is no zero.

Even, given these constant upwellings and contractions-back of the continents, there is no stable centre to the planet. All is movement. Which raises the question of what kind of security can be constructed without a static (immobile and thus unquestionable) foundation (maybe it is in that on-going interrelatedness).

The undertow of many of the artworks meditated upon the specificity of tradition and, in a way, of the past. It was possible to feel in this a critique of the present. And yet, as the knitted sweaters with their statements also seemed to hint, the past is not an alternative future. But there are, precisely, alternative futures. It is not change, in and of itself, that is problematical, but the direction of change. And over that there is some choice. There is something delightful in the fact that Iceland arises from the opening up of the Atlantic. Metaphorically, it speaks of the distance between continental Europe and the USA. Socially, politically, they remain (just about) different possible futures. Small differences perhaps, but enough to make the point that the future is not foretold.

*

If time, then, is the dimension of change – though not in a predetermined direction—space is the dimension of multiplicity (precisely of that multiplicity both of actual presents and of potential futures). It is the dimension of the contemporaneous co-existence of more-than-one-thing. One consequence of this is that space is the dimension of the social (both human and nonhuman). It is not, as it is so often implicitly imagined, a flat surface (so often space is imaginatively equated with the land and the landscape—we travel 'across space'). Rather than as a surface, space may be imagined as a simultaneity of unfinished, on-going, stories. So to walk 'across space' is, in fact, to walk across stories, across a multiplicity of ongoing trajectories. A further consequence is that the multiplicity presented to us by the fact of space requires negotiation. It poses the question of how we are going to live together.

CHAPTER TWO

MATERIAL CULTURE AND SOCIAL SPACE IN THE ICELANDIC TURF HOUSE[1]

ANNA LÍSA RÚNARSDÓTTIR

Introduction

A space can be transformed through the introduction of material culture; the arrangement of objects, their interrelationship and cultural interpretation can have a significant impact on the experience of a space. The sensation of place is linked to the sensible space, as meaning is interwoven with perception. This paper examines the relationship between material culture, physical space and social space, by looking at the case study of the Icelandic turf house, particularly in the period from 1850 until 1966. The key features of this type of vernacular architecture will be explained and social and material changes tied to changes in the architectural form and the social use of space. Using Bourdieu's[2] and Giddens'[3] theoretical framework the analysis will focus on the role of material culture within the space and its impact on social change.

Architecture is a product of culture. However, culture is not a static entity that produces architecture as an end product. The relationship between people and their buildings is an ongoing dialectical process, the building framing the inhabitants' spatial experience, whilst they in turn modify these limits to suit their changing needs. The relationship between

---

[1] The author wishes to extend thanks to the National Museum of Iceland for providing the opportunity to carry out this research as a Dr. Kristján Eldjárn research fellow, to the informants for their warm welcome and willingness to share their memories, to colleagues at the National Museum for helpful comments and assistance, and to participants in the Sensi/able Spaces conference in Reykjavík where an earlier version of this paper was presented.

[2] Bourdieu (1977).

[3] Giddens (1984).

different spaces and the way in which they are defined are reflections of the world-view and ideologies upheld by the society. It needs to be interpreted as a consequence of a wider socio-economic context within which it has to be placed, to be properly understood. Bourdieu claims that:

> ...inhabited space – and above all the house – is the principal locus for the objectification of the generative schemes; and, through the intermediary of the divisions and hierarchies it sets up between things, persons, and practices, this tangible classifying system continuously inculcates and reinforces the taxonomic principles underlying all the arbitrary provisions of this culture.[4]

Ideas then, according to Bourdieu, are embodied within the house and the house acts as a mnemonic device, where symbolic significance is embodied within the house and perceived by its inhabitants. These ideas then help to reproduce the social structure through their very presence in material form in people's everyday lives, that is – through what Bourdieu calls *habitus*. This perception, he claims, takes place on a non-discursive level, that is, the signs are read by the body more than on a conscious level by the mind.

Bourdieu developed these ideas from his fieldwork among the Berber in Algeria, a non-literary society. Change, for example, in the arrival of new artifacts was not a central concern to Bourdieu in this work. The Icelandic turf house, which is the focus of this paper, underwent significant changes in the final stages of its development[5] which need to be addressed. In my research on the development of Tongan architecture[6] I use Bourdieu's theory of practice in conjunction with Giddens' structuration theory that deals with the relationship between social change and social stability.

Giddens argues for a dialectical relationship between the agent (the individual) and the structure (the society), each interacting with the other. The structure is then not only restrictive, but also enabling. The house, as a representation of the structure, thereby becomes a microcosm of the society, its form having been developed by previous generations of society and taken material form at the hands of people. Placing material culture in general in relation to these theories, the object then becomes part of both sides of the model. It becomes a tool for the individual (or the agent) to

---

[4] Bourdieu (1977: 89).
[5] Rúnarsdóttir (2007).
[6] *Ibid.* (2004).

impose their mark on the structure and it also becomes a part of the structure itself, thereby conditioning the ways in which the agent is able to use the tool.

This theoretical framework can prove useful when analyzing the role of material culture in the home within the context of change. After providing an overview of the development of Icelandic vernacular architecture and some of the key aspects of the historical context, a case study of an Icelandic turf farm will be presented. The discussion will focus on two themes: the household and life on the farm and material culture of the home.

Development of Icelandic Architecture

An architectural tradition was brought to Iceland during settlement in the 9th century.[7] It has since had to adapt to the local climate, different environmental resources and the requirements of the society that developed on the island. The research carried out on vernacular architecture in Iceland has mostly concentrated on describing early Icelandic architecture, identifying the different types of buildings, the development of their form and the different construction techniques.[8]

The Icelandic turf farm has developed from the long house; a Northern European tradition brought to the country by the Norse settlers. As their name suggests, turf is one of the main materials used for its construction. Timber is used for the structure, turf is laid to form the thick walls and as a cover for the roof. Sometimes stones are used with the turf for the walls, or as an underlay for the roof. Although the materials have not changed much, the form of the turf farm has not remained the same throughout its history, but rather changed and adapted to a changing context. The long-house was traditionally one large building which housed all the necessary functions. The longitudinal double-pitched roof was supported by free-standing inside posts and covered with turf. The entrance was placed on the longitudinal side of the house. The outer walls were made entirely of turf or turf and stone. Inside, the house may have been divided into two or three parts by light timber walls.

The shape of the long-house gradually changed, as other smaller buildings were annexed to it at the rear. These would at first be the kitchen and the pantry, whereas later other buildings were attached to it as well.

---

[7] See Hannesson (1942); Ólafsson (2004).

[8] See Ágústsson (1987, 1998); Bjarnason (1939); Bruun (1908); Guðmundsson (1889); Hannesson (1942) and Råvad (1918).

With the accumulation of these smaller buildings, the long-house also decreased in size. Thus emerged a new type of construction; a group of smaller houses connected by a central passage, which has been called the passage-farmhouse (*gangabær*). From around 1800 another major change can be noticed in the form with the development of the so called gabled-farm (*burstabær*). This development is often attributed to Guðlaugur Sveinsson who published a treatise on this house type.[9] In this case the facade of the farm was reoriented towards the yard and the annexed buildings at the rear were reintegrated into the main building. However, instead of the being one large building covered with one longitudinal roof, the gabled-farm was a series of juxtaposed smaller buildings, each with its own roof (see figure 1 below). This development was particularly dominant in the south, whereas in the north this new style facade was a front to the old style passage-farmhouse at the back.

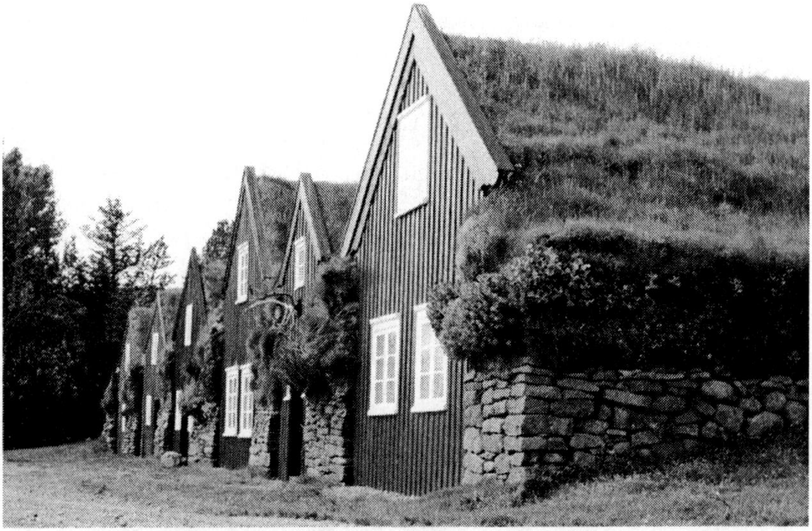

Figure 1: The old turf farm at Bustarfell.

In a country with as long cold and dark winters as Iceland, heating is an important element in house design. Ágústsson[10] argues that the house form changed in an attempt to improve heating efficiency, combined with a

[9] Sveinsson (1791).
[10] Ágústsson (1987, 1998).

decrease in temperature leading to the "mini Ice Age".[11] When firewood was still easily available, in the time of the long house, a big fire was placed in the centre of the floor. As the main and sometimes only room of the house, it was used for all activities. People slept, worked and ate, all in the same room. As a regular supply of firewood became harder to obtain, people gradually began to divide their time between different rooms. The *baðstofa*, originally a bathroom or sauna, was a much smaller room and easier to heat up. First, the head of the household with his wife and children retreated in there for warmth at night. Later they were joined by the older women in the house, leaving the men in the cold *skáli.* Other activities made their way in there as well, until things had come full circle. The *baðstofa* had taken over the role of the *skáli*; once more people slept, worked and ate, all in the same room.

Although Ágústsson emphasises the environmental factors, he briefly mentions other possible causes for change such as the increasing class distinction and the power of the church.[12] Vésteinsson[13] questions the environmental explanation, suggesting that archaeological interpretation would take other factors also into consideration. This echoes the ideas put forward by Rapoport[14] where he states that socio-cultural factors are primary in determining architectural form, whereas environmental factors merely modify the form.

Not much has been written about the development of the Icelandic turf farm beyond the appearance of the *burstabær* and where the *baðstofa* had replaced all the functions of the former *skáli*. It is, however, important to bear in mind that the development continued, and that the wider context of the *skáli* was very different from that of the *baðstofa*. The *baðstofa* was but one of several rooms in the building complex, with other rooms with a specialized function forming other parts of the home. This paper attempts to address the final stage of development, explore it further and within the context of the changes to work and the role of the farm, both as a space and as a social institution.

Methodology

Only a few turf houses remain in Iceland and most of them are preserved within the museum and heritage sector. Five farms were included in this

[11] In Iceland the period from around 1600 to 1900 has been referred to as the "mini Ice Age" due to a change in climate (Þorsteinsson 1966).
[12] Ágústsson (1987: 296–297).
[13] Vésteinsson (2001).
[14] Rapoport (1969).

study, all of whom are part of the National Museum's Historic Buildings Collection, located in the south and east of Iceland.[15] Inhabitation of those houses continued for longer than in many other places, or up until 1966 in one case.

The central tenant of the project is anthropological, relying on informants to supply data on the life of these houses. The number of people that still remember their time living in the houses varies from place to place, from one person to several. Allowances were made in the analysis for the fact that in some cases the informants only remember their childhood in those houses, but key informants are old enough to remember their time as adults. The informants are both men and women that used to live in these houses. The youngest was only six when he moved to a different style house (adjacent to the turf farm), whereas the oldest was in his late twenties when he moved. Most of them were part of the family that owned (or rented) the house, but a few had been there as workers.

Although each farm is different and represents a unique case study, they are all real examples. In this paper a case study will be presented from one of the farms, Bustarfell, in order to illustrate the development. It should by no means be taken to be representative of the whole, but merely to provide an insight into the material. Wherever possible, an attempt will be made at placing the material within a wider context.

Additional data was obtained from the National Archives, in the form of census reports, estate settlements, *etc*. and from the National Museum's historical buildings archive and ethnological collection. This is used both to provide more information about the houses in the study, and to place them within a wider context.

Historical Context

In the time period that my research focuses on, from around 1850 to 1966, many transformations took place in Iceland, both on the economic and social level as well as on the ideological and legislative level. This historical context is important in order to understand the changes that took place in the home. Only some of the key elements that are relevant to this study will be briefly outlined here.[16]

The Icelandic culture and society was based on farming and animal husbandry, with a set of rules and norms that maintained the power of the farmers. Farm servants and workers alike were bound to them, not only

---

[15] National Museum of Iceland (2001).

[16] See also Karlsson (2000) and Kjartansson (2002).

through tradition, but also through legislation. Farm servants were under obligation to have a fixed domicile at a farm for one year at a time, after which at moving days (*fardagar*) they were permitted to move to a different farm. The farmers controlled the whereabouts of the workers and the regulation of their activities. On the other hand the farmer was responsible for the people attached to his farm and had to make sure they were provided for. The social structure was therefore geared towards the maintenance of the power of the rural society.

Iceland was predominantly rural and it was not until the 19th century that urban areas started to develop to any extent. The growth of the fishing industry was one of the key factors in this development. In the same period mechanical advances in farming techniques started to influence the work on the farm, as more and more machines were introduced both for working the land and processing the produce. The difficulty in introducing the new machines was the need to improve the terrain to make it accessible to machines; this took considerable time and effort but was encouraged by the authorities. Co-operatives were also founded and took over some of the tasks that had until then been in the farm's domain. The ideological foundation of the society was affected by all these changes and the legislation followed, allowing workers with no access to land to marry and set up a household in the fishing villages.

Since settlement Iceland had always been partly dependent on foreign trade. Although trade links had been established with several different countries the Danish monopoly of trade (from 1602 until 1787) had a severe effect on the economy with far reaching repercussions on society. Once all trade restrictions had been ... lifted (free trade in 1854), foreign imports started to increase (as well as possibilities for export of produce, and thus income). These economic developments had a significant impact on the availability of material goods and consequently on the material culture of the home. Political changes also marked this period. Following a struggle for independence from the Danish Crown in the 19th century, Iceland was granted autonomy in 1918 and full independence in 1944. The nationalistic discourse and the focus on an Icelandic national identity had an impact on the ideological foundations of society, also affecting the home.

Although this section only briefly touches on the many changes of Icelandic society, its aim is to highlight those factors that are most relevant to the discussion about changes in the Icelandic turf house.

Case Study – the turf farm at Bustarfell

It is within this context that the case study of the turf farm at Bustarfell is presented. As one of the five farms in the study, it is also the one that was inhabited the longest, or until 1966. The farmhouse is quite large and located on a large estate, a few kilometers away from the nearest village. It has been in the same family since 1532. The buildings (see figure 2 below) are composed of a row of six individual houses (rooms), containing three storage areas, two living rooms, and an entrance hall. Both living rooms have an inhabitable upper floor. All rooms but one are connected at the rear by a corridor. At the back of one of the living rooms is the 'new' kitchen, following on from the entrance is another corridor leading to the 'old' kitchen, which leads to the pantry, machine room, cowshed, dung house and an interior well. The main living area in the eastern part of the house is longer than the other houses, and has a large *baðstofa* on the second level, parts of which have been sectioned off into smaller rooms. Underneath the northern part is the guestroom, which was used by the workers. Part of the main house is two rooms accessible only from the outside, a storage (has also been used as a garage) and smithy. In addition to the main building complex, there used to be a set of outhouses, barns, lamb shelters and stables, the traces of which are in some cases still apparent in the surrounding landscape. This description is of the farm's present form, but in the past periodic changes had been carried out with new spaces being added and others being modified as repair and maintenance was carried out.

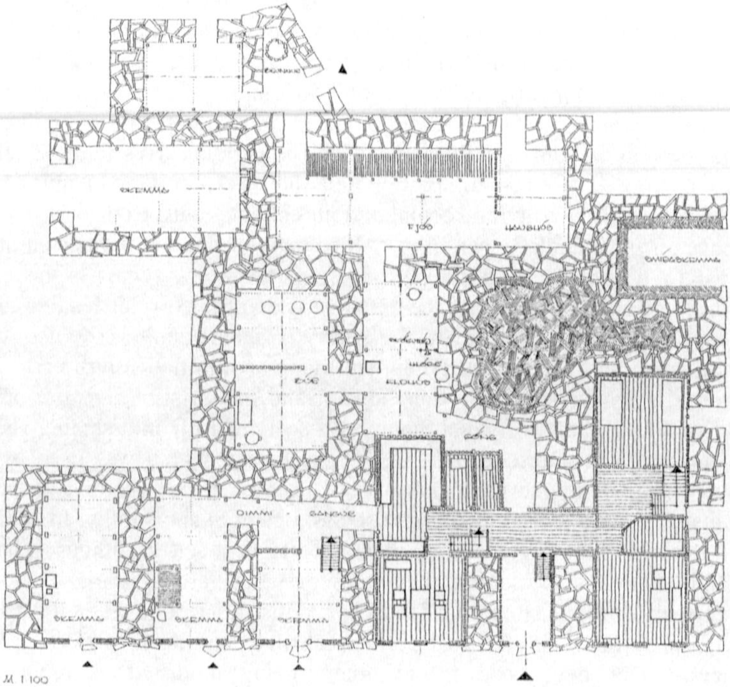

Figure 2: Plan of the turf farm at Bustarfell.

Bustarfell came into the National Museum's care in 1943.[17] When the Museum worked with the family on the renovation of the houses (which had become partly rundown), certain amenities were put into place, which were not part of the normal fixtures of a turf house. A WC and a heating compartment were put in, made with concrete walls for reasons of safety for the boiler. This of course has altered the nature of the space, but made it possible to live in the house until 1966. Most of the living area was wired for electricity. A windmill was set up to generate electricity in 1942, for lighting and to recharge batteries for the radio. By 1944 water and central heating had been installed. A telephone connection was established in 1930, one of the first farms in the area to obtain one. A larger concrete building was constructed near the old farm and the family was able to

[17] National Museum Historic Buildings Collection archive.

move there in 1966. The old farm is still in the care of the National Museum and now houses a local museum.

The number of inhabitants ranged from 20 in 1855, rising to 25 in 1860 and then gradually declining until 1910 when 11 people lived at Bustarfell.[18] The two generations I interviewed explained that towards the end of inhabitation the household composition had been that of a three generation family and some non-family workers, varying at different times of the year. The head of the household was the farmer and his wife, which stayed in the *baðstofa* with their children. Once the children grew up, one daughter married and moved away, whereas the other one married and stayed on. She moved to a separate *baðstofa* with her husband and their five children. In addition to this two workers stayed downstairs in the guestroom. Girls from the area would occasionally come to stay for a period of time to help out in the household, and children would come in the summer from Reykjavík or other towns to participate in the work on the farm.

Guests came frequently,[19] friends/family, travellers and later tourists and foreign anglers. All were very well received and the farmer took his pride in offering his hospitality. The parlor was used to receive guests, and for those staying overnight there was a guestroom (usually occupied by the workers). A large kitchen table was also used for guests, but was used on a daily basis by the family in the last decades before inhabitation ceased. Towards the end more and more people arrived by car (and not on horseback) and even busses with tourists would stop by. All were made to feel welcome and invited in, although this sometimes put additional pressure on the family. For larger events such as weddings, people would also sit upstairs in the *baðstofa*.

At Bustarfell the cooking had usually been done in the main kitchen at the centre of the house, but with the arrival of a new stove around 1875 this changed.[20] As it was small it was at first used within existing rooms. Being fairly small, the new stove was then moved in the winter from the new kitchen up to the main living quarters to make the most of the heat generated. This was on the upper level, thus requiring all water for cooking and dishwashing to be carried up and down the narrow staircase. Later larger stoves were introduced, complete with oven. This created the

[18] Census reports 1855, 1860, 1880, 1890, 1901, 1910, 1920 and 1930.

[19] Very few guesthouses or inns were in operation in Iceland before the 20th century. The custom was for travellers to ask for food and shelter at farms on their way.

[20] Methúsalemsson (1957).

need for a new kitchen. The large kitchen with the fireplace was left intact, but a smaller more 'modern' room was constructed near the storage area.

For entertainment sometimes one person would read aloud a story for the others. Guests provided a particularly welcome change to these gatherings, perhaps travellers needing a place to rest, as they would carry news from other places and perhaps have a few stories to tell. My informants said that before the family had always read from the Bible on Sundays, but that had stopped in their time. People would read books and the farmer made sure that all his children and grandchildren were taught how to read. Once they had a radio that became an important aspect of social life and they would listen eagerly to news and radio plays.

This case study highlights some of the changes that occurred in that time, in the house, the material culture of the home, and in the social use of space. Although I will now discuss each of these changes in turn, they are not independent from each other and have to be viewed in relation to each other and within the general social context. Only then will it be possible to assess the impact these transformations had on the farm and the way of life there.

The Household and Life on the Farm

The number of people in a household was determined by the size of the property and its potential for raising cattle. This would define the labour requirements and thus the household size and composition. All households required a male and female presence, as some of the work was divided according to gender. In cases where the husband died and the wife was allowed to stay on the farm, she would need to hire a man to help with the running of the farm. All the households in this study had some additional servants at the start of the research period (1855 census); although in some cases they were related to the head of the household.

Urbanisation had an impact on the household composition, as workers preferred to set up their own homes in the towns. Alongside the transformation of farming practices, it became increasingly possible for the farmer to work his land with a reduced work force. This led to a change in household composition, the focus being more on the family and less additional servants/workers present. On some of the farms that were included in this study this development coincided with the coming of age of the children of the farmer, making the need for outside workers less, as the responsibility was taken over by the family.

At Bustarfell, the family no longer stayed all together in one living area, but was separated into different rooms. The oldest generation stayed in the main *baðstofa*, whereas the younger couple with their children had moved to another *baðstofa*. The workers then stayed in the guestroom downstairs. This is similar to the development in other turf houses that had large households. The use of the space had then continued its development from where it had come full circle (see discussion above) and everyone stayed together in one room. It had become segmented, with different groups of people enjoying a little privacy.

The changes in household composition would undoubtedly have had an effect on the life on the farm. Although the census shows the change, it does not account for the increase in transient labour that became more important in the 20th century. Great seasonal differences characterised the work of people on the farm. During the summer most of the work would take place outdoors, but during the long winter months the emphasis was on looking after the livestock and working the wool. Workers from urban areas would stay for a few months in the summer, helping out with the additional workload this season represented. It also became common for urban families to send their children to the countryside, to participate in the work and to be looked after during the school holidays.

This data suggests, and is further supported by my informant's accounts, that the nature of the farm (as a home) had begun to change. It developed from being a place to live and work, alongside other people (family and non-family), to being more a place to live with your family, having increased privacy. The nature of the work changed, and the work was increasingly being moved out of the house to collective workplaces.[21] This inevitably affected not only the household, but also the social use of space and role of the house. The social life on the farm in the evenings had been characterised by the gathering in the *baðstofa* of all the members of the household.[22] As the house became more fragmented and new spaces were added, this changed. My research shows that this was not the case in larger households, where norms of privacy had begun to have an impact on living arrangements.[23]

Outside social contact depended very much on the location of the farm. Places along major travel routes received a larger number of visitors, whereas those further off were more isolated. Due to the late development of urban areas in Iceland, and lack of guesthouses or hotels, guests were dependent on the goodwill of farmers for food and shelter on their way. The general rule was that everyone was welcome to stay at any farm on

---

[21] See Ásgeirsson (1988).

[22] See discussion of Ágústsson (1987) above.

[23] See Rúnarsdóttir (2007).

their way, for up to three nights.[24] Farmers usually took pride in offering the best available in food and comfort without charge. Busy farms would often have special guestrooms, whereas in others a bed would be offered up in the *baðstofa* with the family.

Arrival of the car had an impact on several different levels. Travel had never been undertaken lightly and only those with a clear purpose would go. For those that had a car it meant easier access to the growing towns and thus commerce. Some of the cars were also ... fitted with farming machinery, transforming farming practices for those that could afford the initial investment. Once the dairy collectives had been established, the milk truck traveled around the countryside, offering also the possibility of delivering goods from the local merchant. Guests arriving in cars were different from those arriving on horseback. The reception of those arriving after lengthy travel on horses, often in bad weather, was always that of offering the best food and shelter to tired travelers. Women would even stay up most of the night in an attempt to dry the travelers' wet clothes. Visitors arriving in cars had the option of easily continuing their trip, not needing the same kind of attention. In this case a nice coffee would be served, before the guests continued on their way.

Guests, playing an important part in the social life of the home, were no longer perceived in the same way after the arrival of the car and the new means of communication (such as telephone and radio). They were no longer essential for carrying news and their needs were no longer the same when they were able to leave in their car. Although the farmer at Bustarfell continued to greet his guests with a warm welcome, this took place within a transformed context and shows that short visits were more frequent than guests that came to stay for some time.

These changes in social interaction are reflected in the changes to the household composition, the role of servants or workers and the treatment of guests. This highlights the way the requirements of the household had developed, which is expressed through a change in the way the space of the house is used. The family at Bustarfell no longer gathered in the *baðstofa* for all activities. They used different places to sleep, gathered in the living room to listen to the radio and sat next to the stove in the new kitchen during sleepless nights with the children. Furthermore, the space had been altered to suit their needs with new rooms added. In this way the family was able to both change their physical space and their use of the spaces they already had.

---

[24] See Tómasson (2000).

The changes taking place in society, with the increase in urban areas and new developments in the economy had an impact on the way of life on the farm. The social structure was being transformed offering up new possibilities for individuals to lead their lives. The foundation of the turf farm as a place of work was altered, which had consequences for both household composition and social use of space.

Material Culture of the Home

The material culture of the home underwent significant change in the period the research project focuses on (1850-ca. 1950). As discussed above, the increase in foreign trade opened up the availability of goods and offered new options to the home. New types of cooking utensils, kitchen accessories, furniture, and new technology all made their way into the home. Larger, wealthier homes may have felt these changes earlier than smaller, poor ones, and it was by no means a steady uniform development. It was dependent on means, access, location and perceived need.

From open fire cooking, the introduction of stoves made a dramatic change in the way women cooked food. It also changed the way the house was heated, as the new stove created new possibilities in utilising the space. Although the new style kitchen was introduced in some places, the open fire cooking hearth remained the only cooking facility in many of the smaller farms. People also continued to use this in the larger farms alongside the new kitchens, for the big tasks of preparing food for winter in the autumn. The type of environment is dramatically different in these two types of kitchen. The new one had paneled walls and perhaps a wooden floor, whereas the older ones had bare walls covered with soot and a dirt floor. The former was undoubtedly more pleasant and encouraged more extensive use of the space.

Food used to be served in the main living area, the *baðstofa*, where people spent most of their time when in the house. Food was placed in a wooden vessel with a lid on called *askur*, each person having their own and making use of their personal knife and spoon. With the arrival and increase in porcelain tableware, eating patterns changed. With plates and cutlery came the need for a kitchen or dining room table. Initially this would have been set in a parlor and the porcelain being reserved for guests in a formal setting or for very special occasions. Later as it became more easily available, an additional table was placed in the house and the family would eat their every day meals there. This was more common in the new style timber houses that, along with concrete houses, became more common than turf houses in the 20th century.[25] However, this development can also be noticed in turf farms that were inhabited until the middle of the 20th century, such as Bustarfell.

The key tasks on the farm are derived from the main focus on animal husbandry, such as looking after the animals and making sure enough hay was stored for the winter. Furthermore, the produce needed to be processed, such as the wool, milk and meat. The space on the farm was influenced by this and spaces developed as the functions transformed. New equipment, such as the centrifuge, revolutionised the milk processing, which could be carried out much faster than before and thus increasing productivity. Collective freezers also made work easier, as part of the produce no longer required preservation or storage with other means. Knitting machines were also labour saving and took up less space than the traditional loom. The effect of modernisation on the ways of working also reached the spaces within which they were located.

The arrival of electricity signified a major change in the life of the families. During the long dark winter months it changed the way people were able to use their space. During the first years, the families would have locally generated electricity, whereas later the more reliable collective electricity arrived. At the beginning the focus was on having electricity for lighting, perhaps recharging the battery required for the radio, whereas later the electric stove revolutionized the possibilities for cooking. Rather than having to start each morning to rekindle the fire, it was possible to turn on the heat. However, electricity also posed a problem in the humid environment of the turf house, and as a result some farms were destroyed by fire.

When the telephone was introduced from the turn of the 20th century onwards, and later radio from around 1930, it had a significant impact on the social life on the farm, in such a rural context. The social life on the farm was now more oriented towards listening to the radio, both news and radio-plays were popular and the radio mass became a substitute for the long trip to church.

Changes to both work practices and leisure activities had a significant impact on the life in the turf farm. Many of the changes that occurred in the final stage of development of the turf house can be linked to changes in the material culture of the home. The new materials available and the new technologies transformed the context within which the farm had developed. Eating at a table presented a different social interaction from

---

[25] Turf houses account for 52% of housing in Iceland in 1910, but only 27% in 1930 (Hannesson 1942).

when each person ate sitting on his bed. Pleasant rooms for cooking were used in different ways from the old style kitchens. Entertainment and communication was affected by new technologies, which in turn had an impact on social interaction both within the household and with guests.

The requirements of the rural household were no longer the same and in the long run the turf house was no longer able to meet them. The material culture and the spaces within which they were located no longer embodied the same values and the experience of them was altered both through the different social practices that took place there and the meanings attributed to the new objects.

Conclusion

The Icelandic turf farm, as most types of vernacular architecture, shows a strong link to the social aspects of the local community. The form of the house develops as a response to both farming practices and household composition. The social use of space is influenced both by the household composition as well as by developments in material culture of the home. The material culture of the home is dependent on availability of goods, extent of commerce and the economic position of the farm. Although environmental considerations are important in the development of the turf house, as prior research suggests,[26] this research shows[27] that the complexity of an inhabited space is greater than that governed only by those factors.

In this paper I have discussed how the social use of space developed on from the centrality of the *baðstofa*, as a reaction to changes in work practices, changes in household composition and different approaches to guests. Food preparation techniques altered both the form and the use of the space on the farms. Different ways to communicate information, such as the radio and the change in the nature of visitors had a huge impact on the social life on the farm, and thus on the social use of space. The impact of this was significant, new spaces being needed for different functions, but the form was very adaptable in this way (more so perhaps on the larger farms).

Material culture of the home was a significant part of the way of life on the turf farm and social practices were tied to it in many ways. The *habitus* then is not only an embodiment of the norms and values of the

---

[26] E.g. Ágústsson (1987, 1998).

[27] As Rapoport (1969) proposes.

society, but is manifested in the way of life, the work and activities that take place within the home and affects/is affected by the inhabitants.

The introduction of new types of objects, such as porcelain, furniture and radio had an impact on social use of space through a change in social practices. Significant changes in one area are bound to affect another, the *habitus* then being altered by the transformation of the home and the new forms of material culture present. The experience of the space, how the *habitus* is perceived and read, would then have been considerably different if a comparison is made between the start and end points of this research.

The new forms of material culture were not adopted indiscriminately, but follow on from existing social values, such as improving work efficiency. It also shows an interest in improved communication and an aspiration to a better quality of life. The material changes that took place thus contributed to undermining the old way of life that formed the foundation of society where the farm/landowning farmer was at the centre.

Through participation in the development of the turf house and the active use of new types of material culture the inhabitants were able to modify their environment to suit their changing needs. In this way the material culture, be it furniture, stove or electric powered goods, becomes a tool that they, as agents, use in order to influence the *habitus* that conditions their social practices. On one side the development was a response to the changes in the society of which the inhabitants were a part, but on the other side the selection of certain types of material culture shows the individual's possibility to influence the social practices that characterized his life.

The place of the turf house within society also shows a considerable change from being the predominant type to being a rarity. The needs of the society and the lifestyle requirements demanded a different type of housing, made possible by the technological advances that had also made their way into the old style houses. The adaptability of this type of vernacular architecture had reached its limit and now only stands as a reminder of the life and struggle of generations past. The sense of the space in the turf house can only be understood through the experience of all the interrelated spaces of the farm in connection to the wider social and cultural developments that helped to form it. The meanings read on this level through space and the material culture that defines it show how it shapes and is shaped by those that live within that space.

References

Ágústsson, H. (1987) 'Íslenski torfbærinn' In F. Jóhannsson (ed.) *Íslensk Þjóðmenning I, Uppruni og umhverfi* Reykjavík: Bókaútgáfan Þjóðsaga.

—. (1998) *Íslensk bygginararfleifð I, ágrip af húsagerðarsögu 1750–1940* Reykjavík: Húsafriðunarnefnd ríkisins.

Ásgeirsson, Ó. (1988) *Iðnbylting hugarfarsins, Átök um atvinnuþróun á Íslandi 1900-1940* Reykavík: Bókaútgáfa Menningarsjóðs.

Bjarnason, H. (1939) *Húsakostur og híbýlaprýði* Reykjavík: Mál og Menning.

Bourdieu, P. (1977) *Outline of a Theory of Practice* Cambridge: Cambridge University Press.

Bruun, D. (1908) *Gammel Bygningsskik paa de islandske Gaarde: Arkæologiske Undersøgelser*. Kristiania: Grøndahl.

Giddens, A. (1984) *The Constitution of Society* Cambridge: Polity Press.

Guðmundsson, V. (1889) *Privatboligen på Island i Sagatiden, samt delvis i det øvrige Norden* Copenhagen: Andr. Fred. Høst & Søns Forlag.

Hannesson, G, (1942) *Húsagerð á Íslandi* Reykjavík; Prentsmiðjan Edda h.f.

Kjartansson, H.S. (2002) *Ísland á 20. öld* Reykjavík: Sögufélag.

Karlsson, G. (2000) *Iceland's 1100 Years: the History of a Marginal Society* Reykjavík: Mál og Menning.

Methúsalemsson, Me. 1957: *Bustarfell*. Unpublished manuscript.

National Archives of Iceland. *Census reports for 1855, 1860, 1880, 1890, 1901, 1910, 1920 and 1930.*

National Museum Historic Buildings Collection: *Bustarfell* (26). Archive.

National Museum of Iceland/Guðrún Harðardóttir (2001) *National Museum Historic Buildings Collection* Reykjavík: National Museum of Iceland.

Ólafsson, G. (2004) 'Frá skála til gangabæjar: húsagerð á miðöldum' In Á. Björnsson and H. Róbertsdóttir (eds.) *Hlutavelta tímans* Þjóðminjasafn Íslands: Reykjavík.

Rapoport, A. (1969) *House Form and Culture*. New Jersey: Prentice-Hall, Englewood Cliffs.

Råvad, A. J. (1918) *Íslensk húsagerðarlist/Islandsk Architectur* Copenhagen: Andr. Fred. Høst & Søn.

Rúnarsdóttir, A.L. (2004) *Modernity and Tradition: Houses and Material Culture in the Kingdom of Tonga*. Ph.D. thesis, University College London.

—. (2007) *Á tímum torfbæja: efnismenning og híbýlahættir í íslenska torfbænum*. Reykjavík: National Museum of Iceland.

Stenberger, M. (ed.) (1943) *Forntida gårdar i Island*. Copenhagen: Ejnar Munksgaard.

Sveinsson, G. (1791) 'Um Húsa- edr Bæa-Byggingar á Íslandi, sérdeilis smá edr kot-bæa' In *Rit þess konungliga Islenzka Lærdóms-Lista Félags*. **XI.**

Tómasson, Þ. (2000) *Gestir og grónar götur* Reykjavík: Mál og mynd.

Vésteinsson, O. (2001) 'Fornleifar, heimildir og samtíminn' On: http://www.kistan.is/efni.asp?n=503&f=3&u=2. Viewed 20th March 2006.

Þorsteinsson, B. (1966) *Ný Íslandssaga, þjóðveldisöld* Reykjavík: Heimskringla.

Glossary

Askur: wooden eating bowl.

Baðstofa: common room – literally 'bathroom', probably stems from its origin as a sauna.

Burstabær: type of turf farm, characterized by a series of gables facing the yard.

Fardagar: moving days, period when farmhands could take up employment elsewhere.

Gangabær: type of turf farm, characterized by a passageway or corridor connecting several small buildings.

Skáli: main hall of the long-house and sometimes one of the buildings in a gangabær.

CHAPTER THREE

'FELICITOUS SPACE' CONDITIONS FOR SPACES AGAINST AGORAPHOBIA

JULIA NEVÁREZ

Introduction

Agoraphobia has been defined as an urban condition, the fear of the crowd, a kind of malaise and discomfort produced by urbanization density[1] (see figure 3). The movement inhibition that characterizes the fear of the outside and the crowd is for the agoraphobic a symptom triggered by the urban condition. The psychological and emotional challenges involved in urban living, its heterogeneity, density and excess have been identified as producing at times social and psychological disorganization. What soothes or reduces this fear, the remedy or therapy for this generalized urban anxiety can be found in spaces that provide calm, refuge or what Bachelard[2] calls 'felicitous space'. In this chapter, I discuss different ways in which agoraphobia has been understood from the psychological perspective on agoraphobia as a pathology; as it has been related to urbanity, more precisely urban modernization; from a feminist perspective as the ideological manufacturing of fear to the outside and its consequent spatial limitation for women, reduced to the safety of the home; to contemporary fears of terrorism that might hinder a fuller participation in the public sphere represented by the fear to the openness cities provide. Agoraphobia to a certain extent is based on a closed up notion of space as a container, where no relationship to other spaces is recognized. I then proceed to emphasize the importance of the spaces in the city and propose—as one of the main arguments in this text – that in order to fully unpack the conditions that trigger agoraphobia at different

[1] Carter (2002).
[2] Bachelard (1969).

levels and the possibilities for its soothing, a better understanding of the concept of space, one that assumes it as relational, is necessary. I then discuss different approaches to the study of space through rhythmanalysis, *habitus* and hodological space to present and discuss the notion of 'felicitous space'. The last section includes a typology of places that could be considered to promote or to possess the qualities of felicitous spaces that invite us to the outside. This exemplar typology of spaces we love or are attracted to, is placed within the context of the trajectories that connect the spaces we inhabit and give meaning to our everyday lives. Space is then defined not so much in terms of the traditional view that understands it as a vacuum or a container but rather as social in nature related to other spaces, contexts, situations and experiences.

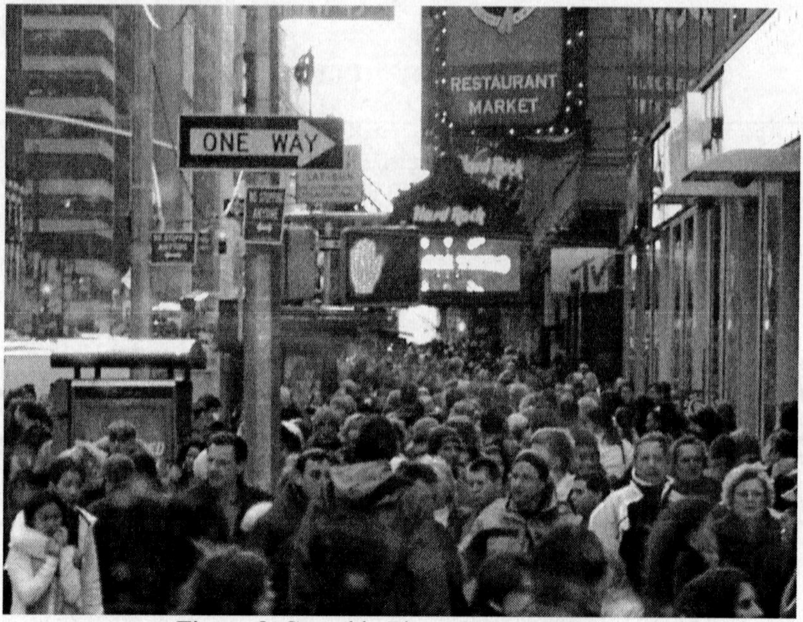

Figure 3: Crowd in Times Square, NYC.

Borrowing from urban studies' notion of agoraphobia, Foucault's space as relational, Bourdieu's *habitus*, Lefebvre's rhythmanalysis, and Bachelard's felicitous spaces, conform the conceptual trajectory used to fit in spaces that offer alternative meanings and experiences, spaces seized by the imagination. Daydreams and spaces identified with our solitude, the spaces that offer momentary glances into surprise and unexpectedness, and the urban uncanny are some of the ideas that will guide the selection of urban spaces presented in the typology of felicitous spaces. To understand spaces that offer the possibility for the development of the imagination is what people perceive to be spaces where they could be engaged in ways that provide comfort, refuge, encourage exploration and even dissent. These spaces can be reconstituted and reformulated through visual perception and movement, in the connectedness derived from journeys as meaning emerge through the spaces of everyday life. Accordingly, the spatial trajectories that frame the meanings of visually reading a landscape, is coherent with a notion of the relatedness of different spaces that signify and produce meaning for urban space.

Representations of agoraphobia: Agora phobia (digitalis)

At Eyebeam, a gallery in Chelsea, NYC that focuses on the intersections of art and technology there was an exhibit in 2003 entitled: *Agora phobia (digitalis)* by Karen Lancel. At that exhibit, the artist placed a sculpture that looked like a bubble, where people could go inside and in a computer terminal answer a few questions, such as:

> What is a safe place for you?
> How would you describe an 'unsafe place'?
> When and where do you feel isolated?
> How do you maintain control over your space?

One person wrote in Spanish inside the bubble at the computer terminal (all the entries where available to whomever ventured inside):

> Ponlo ya.
> Que me ahogo en la calle
> Que hay demasiado espacio y la
> atmosfera esta altisima.
> El arte es un sentimiento.
> Tambien lo es la agorafobia y
> desde luego lo que propicia su cura…
> un objeto de proteccion.
>
> esta obra de arte es un gran condon
> que actua contra la esterilidad
> fructifera de los espacios abiertos a
> primavera.

In translation:

> Put it on already.
> That I drown in the street
> That there is far too much space and the
> atmosphere is so high.
> Art is a feeling
> So is agoraphobia and
> obviously, what brings about its cure…
> an object of protection.
>
> this art work is a great condom
> that acts against the fruitful
> sterility of spaces open to
> spring

Other answers to the questions at the computer terminal inside the bubble were:

> The speed of the unknown while traversing
> sites and paths heals my agoraphobia…
> Then I need a rest with myself to
> regroup and develop the courage to the open.

These are both reactions to the contemporary definition of agoraphobia which Carter defines as movement inhibition. According to Carter, immobility is produced "not by lack of directions, but by an excess of them".[3] We could consider living in a society where freedom is largely measured by the number of choices one has access to, as something at times problematic. There seems to be a dizziness at times triggered by having the opportunity to choose from so many options.

A concept developed from the Greek word agora for market or market place, agoraphobia literally means fear of open space. Coined first by the German psychiatrist Westphal in 1871, he referred to agoraphobia as: "the impossibility of walking through certain streets or squares. Or possibility of so doing only with resultant dread of anxiety",[4] in other words, the fear of going out. The word used for this condition in german was 'platzangst', in French 'peur des espaces', 'horror du vide', and in English 'street fear' in addition to agoraphobia.

The anxiety or discomfort of walking across a vast empty place has also been associated with agoraphobia. According to Camillo Sitte

[3] Carter (2002: 17).
[4] Hurst Vose (1986: 36).

agoraphobia and claustrophobia are both considered what he calls topophobias: "a condition seriously exacerbated by the dimensions of the space, especially "when there seemed to be no boundary to the visual field".[5] Bachelard identifies in Supervielle's writing, a juxtaposition of claustrophobia and agoraphobia when he writes: "Trop d'espace nous etouffe beaucoup plus s'il n'y en avait pas assez". (Too much space smothers us much more than if there were not enough). Bachelard continues stating Supervielle's familiarity with "exterior dizziness" and "interior immensity". Thus, according to Bachelard "the two spaces of inside and outside exchange their dizziness".[6] For an agoraphobic, the prison could be the outside. To which Hurst Vose so eloquently declared: "No fear without the void, no calm without the appearance of a semblance of protection".[7]

From a psychiatric perspective, agoraphobia is considered a pathologic condition symptomatic of the fear of the crowd. For Hurst Vose the situations most cited by agoraphobes as triggers to their condition are:

> streets, shops, churches, cinemas, theaters, traveling tubes, buses, trains, or any other form of travel where an easy exit is not always possible, crossing bridges and going through tunnels. Shopping is a particular nightmare, with supermarkets and department stores presenting a lot of problems. Coaches, ships and airplanes, and close spaces such as lifts are also feared, although cars do not appear to present the same amount of problems.[8]

Maybe because cars even though enclosed spaces, also provide movement and the control through volition imposed on a machine.

Agoraphobia has had an etymological trajectory that is based on a genealogy of spatial fears. Different perspectives—psychological, urban, feminist—have offered nuanced interpretations of agoraphobia, all involved a similar thread: the notion of the fear to the outside, of movement inhibition.

Urban mobility and its discontents

The possible hesitation that accompanies venturing outside, characterizes what critics of modernity consider agoraphobia to be: a symptom of urban estrangement. Moreover, agoraphobia has been identified as a spatial and

---

[5] As quoted in Hurst Vose 1986: 14.
[6] Bachelard (1969: 221).
[7] Hurst Vose (1986: 16).
[8] *Ibid.* 206.

urban disease, due to the presumed debilitating atmosphere of big towns, 'malaria urbana'.[9] Despite the fact that agoraphobia might be understood as fear of open spaces, fear of crowds or movement inhibition, it is evident that all those definitions share a common element found in urban living. The density of urban areas, the reliance on interest groups rather than relationships based on kinship, the accelerated pace of living are all characteristic of urban areas. Vidler considers agoraphobia to be a "historical symptom of environmental change, one of a constellation of anxieties associated with urban modernization".[10]

The labyrinthic spaces of the city have been construed as a source of modern anxiety such as phobia and alienation. What Vidler calls the "generalized fear of the metropolis" points towards the 'surmenage mental' of modern civilization. The spatial pathology of the city is also found in the profound impact of simultaneous and diverse stimuli to which Simmel characterized urban living. The space of the city causes psychological alienation due to psychic overload.

Also viewed as an urban condition, feminist psychologists include in the definition of agoraphobia not only individual pathology but also its relation to women's experiences in general. In a patriarchal society where emphasis is placed on home been the place for the woman, the presence of men largely dominates public space. Moreover, the ideology of women's fear of crime and harassment in public space tends to justify an agoraphobic reaction on them. According to Deutsche "The phobic woman may try to define, and stay within, what she considers a zone of safety".[11]

However, the permanent emergency—and I would add anxiety—that according to Miles[12] characterizes contemporary urban living under the threats of lets say terrorism and a myriad of other fears, could be considered as constitutive of—to a certain extent—agoraphobic times, projecting a future of voids. The contradictory message delivered from authorities through the media to New Yorkers after September 11th was "Be alert but stay calmed". Moreover, yet another interpretation of agoraphobia could be applied to the fear of social openness characteristic of the exposure to the heterogeneous quality of urban living and/or the openness to the ideas that urban living welcomes, of a cosmopolitan way of life in its broadest sense. This is a similar point to the one made by the group of architects in Sarajevo who coined the term "urbicide", referring

[9] Vidler (1994).
[10] *Ibid*. 25.
[11] Deutsche (1986: 325).
[12] Miles (2004).

to the process of destruction of buildings, institutions and ethnic cleansing that took place in Sarajevo. Their argument is that not only buildings and people, but also urban living and the institutions that represent its heterogeneity were attacked and destroyed.[13]

Deutsche introduces the notion of ideological agoraphobia to point at the fear of "engaging in the challenge that democratically constituted spaces (and their publics) represent".[14] The fear to participate in the public sphere and to expose oneself to the diverse and often conflictive points of views debated there are according to Deutsche the ideological agoraphobia that:

> If so, all those books and articles mourning the lost beginning of the public sphere are not mere responses to the fact that the public is not here. Taking the form of what Keenan calls misrecognition, they are, rather, panicked reactions to the openness and indeterminacy of the democratic public as a phantom—a kind of agoraphobic behavior adopted in the face of a public space that has a loss at its beginning.[15]

Freud's notion of the uncanny seems appropriate in this regard, to help us establish yet another angle from which to discuss the agoraphobic symptom and its possible cure. For, if the uncanny is the strangely familiar,[16] one way to overcome agoraphobia is to discover the protection of the familiar in spaces and places. According to Freud ".... the uncanny is that species of the frightening that goes back to what was once well known and had long been familiar".[17] This could be an element of the urban environment that could trigger agoraphobia, that which stopped been familiar arousing feelings of strangeness. The urban uncanny corresponds to the unhomely and its symptom. The opposite could be its cure: homely. How could the 'homely' be also placed outside? To all these related applications of agoraphobia to contemporary times, Bachelard's notion of 'felicitous spaces' is an important and possible consideration for spaces against agoraphobia. However, there is yet a more comprehensive perspective on space that speaks for the use we have of urban environments in everyday life and that adds a temporal component to the analysis of the actual experience of space. The solution to the agoraphobic might not be contained in one single space but a feeling of connectedness and relatedness to the many spaces of the polis.

---

[13] Coward (2004).
[14] Deutsche (1986: 185).
[15] Deutsche (1986: 325).
[16] Freud (2003).
[17] *Ibid*. 124.

Spaces and trajectories: Space as relational, rhythmanalysis, habitus, and hodological space.

George Sand wrote: "what is more beautiful than the road? It is the symbol and image of an active, varied life."[18]

As relational and connected, how do we conceptualize the spaces where everyday life happens as those providing opportunities against agoraphobia, for 'felicitous spaces'? A more active and broader conceptualization of space seems necessary, one that is outside the more traditional conception of space as an abstract, empty container and/or void disconnected from any kind of context. Spaces that relate to each other, connected by use and meaning in addition to social, political and economic context offer a richer and fuller understanding of space, conceptually. Foucault[19] provides what I consider to be, one of the most challenging and appropriate definitions of space as relational. In his definition space is a skein of relationships, networks of relations among sites. It is indicated in this definition the need for an understanding of space that links, for a notion of space that is in fact connected. Even though he also speaks for space where simultaneity and juxtaposition among things, people and/or ideas happen, implying an absence of relations, he does state that it is only in its appearance that space presents itself as disconnected and relation-less. The apparent simultaneity and juxtaposition of disconnected events and spaces is something that could be identified in the apparent display, not in the 'actual' set of relations that distinguishes contemporary spaces. The relationships among spaces—at various levels—is what constitutes space and gives it meaning.

At the level of the body and culture, Bourdieu's[20] notion of *habitus* structures the connection between different spaces, culture, and our bodies, ourselves. Through the *habitus*, we maneuver through spaces mediated by culture but also embody culture in our practices, in our uses of everyday spaces. By linking the spaces of everyday life, we engage in the agency of challenging container or void space, with such dichotomies as inside, outside, far and near, that even though might be adequate analytical tools, are opaque. They do not address the in-between areas where inside and outside, for instance, might be confounded. At the analytical level of how urban space is produced Lefebvre[21] offers the triad

[18] Cited in Bachelard (1958: 11).
[19] Foucault (1986).
[20] Bourdieu (1990).
[21] Lefebvre (1984).

of: representations of space, representational space and spatial practice. This triad attempts to offer an analytical tool by which the complexity of urban space could be better appraised. It is within this triad that ideologies, practices and discourses about space can be examined to offer the necessary differentiation of the social nature of spatial meanings. The linking of space generates the movement where Bourdieu's *habitus* and Lefebvre's analytical triad challenge notions of space that rely on static, dichotomous and empty notions, a space basically treated as inert matter. Providing this kind of life to the understanding of space could help to articulate 'felicitous spaces' that are linked through rhythms and paths and that provide a more accurate depiction of the way space is in fact experienced.

Even Sartre suggests the study of space as hodological: the analysis of paths. For Sartre "Rooms, doors and windows are never quite entirely passive realities…[they are] instrumental structures and products of direct activity".[22] Space is instrumental and so is the ability to link spaces, to understand everyday trajectories and journeys that provide social meaning to space, by understanding it as multiple sites, connected through social and cultural practices. The moments and situations that emerge from this linking result from the movement through space. It also emerges from the desire to understand what is lacking in space, who is invisible, who is present, what is represented in space as we engage in the present forcefully and critically, with others. Multiplicity of moments, situations, journeys, and trajectories, are the conditions for emerging instances of 'felicitous spaces'. The way spaces could be linked – behaviourally, psychologically, experientially—is through movement. In the city one of the most common forms of movement is to walk.

Flaneur/Vagabond/Walker the discovery of 'felicitous space'

Benjamin, slow walking is the art of the *flaneur*, a wonderer who challenges the rigid grid or what Carter calls the 'lie of the land', the artificial and mechanistic compartmentalization imposed on urban space. A different category unto itself and differentiated from the bourgeois *flaneur* is the nomadic route of the vagabond (usually of a lower class) in the labyrinthic spaces of the city. The analysis of these routes is what Bachelard—as does Lefebvre—considers the possibility of using rhythm analysis as an analytical tool to understand the rhythms and "reconcile and

[22] Sartre (1968: 153).

lighten the ambivalences that psychoanalysts find in the disturbed psyche".[23]

What are the conditions that generate 'felicitous spaces'? According to Bachelard: "Space calls for action, and before action the imagination is at work".[24] That imaginary produced in space rests in "the domain of intimacy, in which psychic weight is dominant".[25] Bachelard is interested in examining, 'felicitous space' or eulogized space: Indeed, the images I want to examine are the quite simple images of felicitous space. In this orientation, these investigations would deserve to be called topophilia. They seek to determine the human value of the sorts of space that may be grasped that may be defended against adverse forces, the spaces we love. "For diverse reasons, and with the differences entailed by poetic shadings, this is eulogized space".[26] He continues, "…we should have to undertake a topoanalysis of all the space that has invited us to come out of ourselves".[27]

Bachelard's 'felicitous spaces' could also be interpreted as the space that invites us to come out, thereby offering solutions and possibilities against agoraphobia. In his poetics Bachelard transforms everyday objects and spaces into psychic charged objects. The depth of sentiment and emotion with which he treats and describes shells, corners, the house and other sites and objects evoke a tender and warm feeling. The recognition of those feelings as well as the gratitude found in the comfort of familiar ideas and experiences pose to a large extent -and even though similar to the way other people experience space- unique contents. Phenomenological in essence, Bachelard's poetics of space does not consider the political ramifications of a very intimate experience of space where negative experiences such as abuse and/or violence might take place. This, however, does not seem to be his experience nor the focus of his project. Rather, space as well as imagination and poetry serve to depict the psychological dimension; the way life is experienced in and through space, in its fullest positive capacity. Only his notion of transubjectivity seems to bridge the gap of individual experience, still remains apolitical and ahistorical. Consequent with his undertaking is to search instead for that which is similar in experience, rather than that which might be considered conflictive and/or contested.

---

[23] Bachelard (1958: 65).
[24] *Ibid.* 12.
[25] *Ibid.* 12.
[26] *Ibid* xxxi.
[27] *Ibid.* 11.

Bachelard presents to us the notion of alveoli space and as such make us assume the space of that which is somehow, nested, cocooned or sheltered, where the intensity of intimacy and its psychic energy is concentrated, and of something that might have left a trace. For Bachelard we need both; retreat and expansion, simplicity and magnificence, and he uses the metaphor of the chrysalis to represent unavoidable tensions such as the inside/outside. To illustrate the tension of dualities Bachelard states, "To sleep well we do not need to sleep in a large room, and to work well we do not have to work in a den. But to dream of a poem, then write it, we need both".[28]

The felicitous spaces I aim to introduce belong to a different set of categories and should be understood as part of the paths, the journeys and trajectories of everyday life in a city. They are related to the urban experience and indicate degrees of different qualities that offer an invitation to the outside. The imagination is, according to Bachelard, the faculty of producing images. Accordingly, the following images described are also enmeshed within the perspective of the photographer, an image within an image. They represent a quite personal sampling of the kinds of felicitous spaces that can be found in the city, that offer comfort, beauty, novelty, picturesqueness, variety and unexpectedness in events. These are linked spaces through the movement of walks, riding bicycle and other means of transportation basic for urban living. For instance, the Sheep Meadows in Central Park offer a sense of relaxation basically triggered by the openness and prospect of the long lawn and the silhouette of the city's skyline as backdrop. Even though serene, it is not the kind of landscape that induces loneliness, since there is always traffic nearby. The Mall, also in Central Park, offers an elegant sight of landscaped trees, that invite to stroll. The Boathouse in Central Park as well, provides an imagery that satisfy our aesthetic sensibilities if and when those seek a majestic serenity as does the bridge by the lake. The solitude that could be found in those landscapes is not one that alienates but rather is complemented with that of the designed nature that Central Park offers as a refuge from the ribbons of asphalt and concrete that border its perimeter. Art galleries also offer a different kind of felicitous space in the metropolis. These could be considered image machines that stage art contents defined as broadly as possibilities for the imagination exist. Images trigger daydreams.

A different and sometimes controversial kind of art is the one that appears throughout the city, in the walls of unofficial art establishments in many neighborhoods. These could be considered spaces of the unexpected

---

[28] *Ibid*. 65.

when while strolling or walking through the city an image surprises our senses. Those produced in New York City by the self-proclaimed street philosopher DeLa Vega where in a chalk drawing on the sidewalk a fish jumps from a large bowl to a small one, as if conveying the blind ambition that sometimes guide urbanites or his *Homage to Picasso* where *The Guernica* is reproduced on a wall in El Barrio, East Harlem. These are momentary glances into the unexpected, into an image that is accidentally found.

Other kinds of felicitous spaces are those that even though invite us to the outside, they offer refuge. Not addressed by Bachelard, but certainly discussed in contemporary cultural criticism, the feeling of protection that home offers is also described as the site of 'capsularization', the retreat to the private sphere as a reaction to the increasing threats offered by public space, such as terrorism and/or crime. Within this perspective, the inside as a capsule could be considered the epicenter of the neoliberal individual. The individual required to function under current advanced capitalism: self-sufficient, efficient and entrepreneurial. Spaces that offer refuge and comfort outside the home are a different kind of option for the urban malaise of agoraphobia.

In the New York Public Library one can find spaces of exploration, where the curiosity and the satisfaction of intellectual engagement could be found. Felicitous spaces could also be constituted by dissent, where voices of difference and contestation can be heard such as the Sheep Meadows where the group Billionaires for Bush satirized the 2004 Republican National Convention. They also move around the city and could have been found in demonstrations and other public spaces of the city voicing their resentment, frustration and complaints.

In the city, the infrastructure of the buildings could offer both, the uncanny and the homely. The skyline could offer a sense of familiarity that is comforting and at some other times it could resemble the specter that triggers the agoraphobic reaction in the void of an empty city.

Conclusions: Conditions for spaces against agoraphobia

The conditions for felicitous spaces or spaces against agoraphobia require an orchestration of many factors. They require an economic base that is more interested in communal and collective experience, but also where individual experiences can find its own mode of expression. The oppressive atmosphere of surveillance, fear and control that undermines urban living, especially for the protection of capital in different circuits and its ideologies, seems misguided. Critically, felicitous spaces could provide the opportunity to develop alternative readings of the urban fabric and the social text embedded in it by compelling us to develop criteria under which we could claim our right for the kind of city that offers an invitation to come out and live it. Challenging the times in which we are living, especially in US, an artist mentioned some of the following to offer possibilities for a critical engagement: discipline, allowing disagreement, ambition and desire, and I would add playfulness and creativity. Many of these could be applied as conditions necessary for felicitous spaces that contribute to creatively manage the dialectics and tensions between inside and outside, movement and inertia, crowd and void, openness and closeness, fear and safety, danger and protection, where fitting in is at its most phenomenological and psychic level such as the one addressed by Bachelard in his poetics of space.

Agoraphobia can happen at many different levels: psychological, urban, spatial and ideological. The negation of the possibility of openness, movement inhibition based on fear is something becoming increasingly generalized and intensified with the fear of terrorism and crime that permeates most urban, suburban and even the few non-urban areas. The notion of felicitous space, therefore should provide the opportunity to articulate options and possibilities that generate spaces and atmospheres against agoraphobia or its sister claustrophobia. Spaces that invite going outside.

Art can play a major role in this process by opening possibilities for experiencing spaces and the city differently, where creativity is encouraged rather than repressed, even though both of these attempts tend to trigger creative responses anyway. However, nourished creativity versus creativity that spurs from repression involves a different set of social relations. Both require the need for solid member kindred spirit groups, but the search for those seem more difficult under a repressed atmosphere of fear that tends to bring about distrust. Even though Freud states that, "It may be said that those who are happy never fantasize – only the dissatisfied",[29] to work collectively and for collective change might require better means to articulate dissatisfaction.

Miles[30] also considers the role of art and urbanity to be important. He argues for a new urbanism based on Gablik's sense of connectedness. That idea is similar to the previous discussion of paths and trajectories, spatially speaking. The transformation of the city, the function of art in this transformation and in the appearance of the city should, according the

---

[29] Freud (2003: 28).

[30] Miles (1987).

Miles seek to offer an ecologically responsible and communitarian society. Most likely, this will also be a contribution to the creation of felicitous spaces against agoraphobia where the invitation to come outside is not considered an insurmountable task filtered by fears, but an ease into the spaces of others. The acknowledgement that space is relational can help understand the social and political dimensions of the society in which been inside should not compete with been outside, and much less to do so based on fear.

References

Bachelard, G. (1969) *The poetics of space* Boston: Beacon.

Bourdieu, P. (1990) *Outline of a theory of practice* Cambridge: Cambridge University Press.

Carter, P. (2002) *Repressed spaces: the poetics of agoraphobia* London: Reaktion.

Coward, M. (2004) 'Urbicide in Bosnia' In S. Graham (ed.) *Cities, war and terrorism: Towards an urban geopolitics* London: Blackwell.

Deutsche, R. (1986) *Evictions: Art and spatial politics* Cambridge: MIT Press.

Foucault, M. (1986) 'Of other spaces' In *Diacritics* **16**: 22–27.

Freud, S. (2003) *The uncanny* London: Penguin Books.

Hurst Vose, R. (1986) *Agoraphobia* London: Faber & Faber.

Lefebvre, H. (1984) *The production of space* Oxord: Blackwell.

—. (2004) *Rhythmanalysis: Space, time and everyday life* London: Continuum.

Miles, M. (2004) *Urban Avant-gardes: Art, architecture and change* London: Routledge.

—. (1997) *Art, space and the city: Public art and urban futures* London: Routledge.

Sartre, J.P. (1968) *Search for a method* Vintage Books, New York.

Vidler, A. (1992) *The architectural uncanny: Essays in the Modern unhomely* Cambridge: The MIT Press.

Vidler, A. (1994) 'Psychopathologies of Modern Space: Metropolitan fear from agoraphobia to estrangement' In M. Roth (ed.) *Rediscovering history: Culture, politics and the psyche* Stanford: Stanford University Press.

CHAPTER FOUR

OPENING UP THE OPEN SPACES THROUGH SPACE SYNTAX

JULIENNE HANSON, REEM ZAKO, HELEN BENDON AND JESSICA THOM

Introduction

In many parts of the world, housing and neighbourhood developments of the late twentieth and early twenty first centuries have been shaped by the belief that it is important to create communal open spaces that can provide a physical focus for local communities. This has given rise to a range of modern housing typologies that are based on an ethos of building community through design, using concepts such as a 'privacy gradient', the separation of pedestrian and vehicular movement and the notion of 'enclosure', amongst others, to shape the public open space of the development. This paper presents early findings from a collaboration between architects and artists that set out to examine the contribution of such public spaces to community awareness in thirty English city-centre housing estates and residential areas with differing housing morphologies, built between the middle of the nineteenth century and the early twenty-first century. As a manifestation of the complex social relationships that take place within the domain of public space, we suggest that the above concepts and their physical realisation need to be complemented by careful consideration of social notions such as the constitutedness of the spaces, co-presence and natural surveillance. The methodology used is based on a detailed analysis of the built forms and spatial layouts of the estates, combined with a 'space syntax' analysis of the interconnectedness and accessibility of the public realm within each estate, representations that show the extent of visual fields from key public open spaces and an appraisal of the types of physical boundaries surrounding each of these open spaces and the openings that give onto them. The juxtaposition of these different analyses provides insight into the interfaces maintained spatially between residents and non-residents and the possibilities that are thereby created for mutual awareness and encounter. These more scientific notions about how to address the public open spaces of residential areas in towns and cities have been complemented by the production of works of art by two artists in residence, that demonstrate both common ground in interpreting the urban fabric and complementary interests in the construction of the boundary between public and private space.

Background to the Project

This collaborative project between artists and architects is set within the context of a large, government-funded, university-led and industry-partnered research consortium, known as VivaCity. VivaCity is an ambitious, £2.75 million, five-year research consortium (2003-2008) developed in response to the UK Government's need to promote more sustainable urban environments. Focusing on conflict resolution in the twenty-four hour city, it brings together more than thirty university academics with over one hundred construction industry collaborators, to deliver practical tools and resources to support sustainable and socially responsible urban design decision-making. By enabling designers, planners and developers to create innovative solutions, these tools and resources will address issues of crime, economic vitality, the urban environment and social inclusion, thus improving the quality of urban life.

One of the work packages within the consortium directly addresses the design of residential districts over the course of the last one hundred and fifty years. This work is being carried out by the two architect-authors of the paper, Julienne and Reem. Julienne is an architect best-known for her contribution to the theories and techniques for the analysis of spatial configurations known as 'space syntax', which have been developed with Bill Hillier and other colleagues at the Bartlett, University College London, since the mid 1970s.[1] From the outset, the aim of space syntax was to provide tools to help architects understand and predict the likely social consequences of their designs. The fundamental idea behind space syntax is that architectural designs for buildings and urban places can be broken down into their component spaces, which are then represented as maps and graphs that describe the relative connectivity and integration of those spaces in such a way that they can be analysed as networks of route choices. By so doing, it becomes possible to quantify and describe how easily navigable any building or urban area is, and to predict the social

[1] Hillier and Hanson (1984); Hillier (1996).

effects of spatial layouts such as their relative busyness or quietness in terms of pedestrian and vehicular movement, or their vulnerability to crime and anti-social behaviour. Reem, also an architect by profession, has worked with Julienne for the past ten years, on a series of research projects that use space syntax to understand the relationship between house form and culture, the latest of these being the VivaCity project that is reported here.

However, VivaCity also has two artists in residence attached to the consortium, Helen and Jessica. Their remit is to foster the public understanding of science, including the different ways of seeing and interpreting the city among academics and by the general public. Helen is a UK-based artist working predominantly in video and photography. Her work was selected for New Contemporaries in 1998 and since then her films and video works have been shown at international art exhibitions, festivals and screening events. Helen's research and practice focuses on a visual exploration of the relationship between physical and psychological space and is anchored around intimate narratives, often understated or tacitly conveyed. Jessica, a recent graduate from the MA in Photography at the Royal College of Art, joined Helen on the nine-month long, part-time residency within the VivaCity consortium, in order to examine the complexities of the urban environment in which people live. By removing information from photographs, and compiling multiple viewpoints, she highlights the diversity of the details that are found in inner city neighbourhoods. Her work documents why the city looks the way it does and how the choices and demands made by countless numbers of anonymous people are reflected and imprinted on the environment, whether on a grand scale or simply by a line worn in the grass from many people walking the same short cut.

All four co-authors collaborated during the autumn of 2005 and the spring and summer of 2006, to share and compare architectural and artistic representations of the three cities that the VivaCity team as a whole is researching, London, Manchester in the north-west of England and Sheffield, in the north-east. The relationship with Jessica involved sharing the architects' extensive archive of photographs of residential areas and housing estates from the cities under consideration. That with Helen was more abstract and formal in that she was stimulated by theoretical debates that involved all four co-authors to explore themes held in common with space syntax analysis, such as natural surveillance. The paper presents some early findings from this challenging and innovative interdisciplinary collaboration. The narrative is predominantly from an architectural perspective.

Residential Districts

An important constituent of any city is its residential districts. For most of urban history, housing has been the dominant land use and the most important ingredient in giving a distinctive shape, form, appearance and character to cities throughout the world. The architectural research concentrates on capturing and describing the design and layout of thirty housing estates and residential developments in three cities, that have very different origins and place histories.

Clerkenwell is a mixed use 'urban village' close to the heart of the City, London's financial centre, that dates back well over a thousand years and has evolved in a piecemeal, organic and unplanned way. Hulme is a 19th century, working class industrial suburb close to the centre of Manchester, that grew up rapidly in the heyday of the cotton industry as a mixed industrial district of large factories/warehouses and small terraced houses, and has since been redeveloped twice, during the 1960s and the 1990s. The Devonshire Quarter of Sheffield was also developed as a residential and commercial area during the 19th century, based on traditional light industries such as cutlery manufacture, but it has recently taken on a new lease of life as the focus of a new 'urban village' with residential and mixed uses based on specialist shopping, leisure and entertainment facilities.

England has seen a huge amount of architectural experimentation in housing during the last one hundred and fifty years, and the housing heritage of our three study areas has crystallised this experimentation in its built form and fabric, resulting in a 'living laboratory' in which to study the relationship between housing and local communities. Today, housing is once again at the top of the UK government's agenda as the whole country is currently experiencing a massive housing shortage. Housing typology is once more under the microscope as both the public sector and the private sector struggle to reinterpret traditional British housing typologies and to invent new ones that are appropriate to life in the twenty-first century, and it is this debate that the architectural/artistic collaboration is, in various ways, attempting to inform.

The Urban Block

All cities are made up out of an arrangement of streets and urban blocks (islands, insulae), which are further subdivided into plots (or lots) on which buildings stand, and residential districts are no exception. These are the basic elements out of which the urban fabric is constructed. Over the last one hundred and fifty years, there have been several dramatic transformations in the residential urban fabric as a result of the architectural experimentation previously mentioned, and it is these transformations in the built form and fabric of English residential districts that we are attempting to represent, analyse and characterise in our research.

Architects describe this change as a shift in the figure/ground relationship between buildings and public space that together make up the urban tissue. This shift is usually portrayed as one from a city carved out of the solid, where buildings predominate and the streets and squares are lined and defined by building facades so that they look almost as if they have been chiselled out of a solid three dimensional block, to a city in which isolated buildings stand marooned like pavilions floating in an open landscape and where the streets and roads also bear very little relationship to building *façades*.[2] Recently, however, the pendulum has swung back towards a denser urban fabric. This has been accompanied by a resurgence of interest in the design of the urban block, see figure 4.

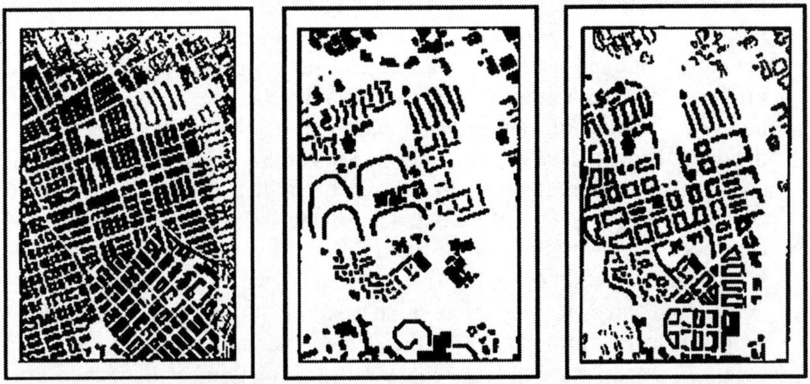

Figure 4: Figure/ground plans of Hulme, Manchester showing the different development densities, at: a) The end of the 19th century: A City Carved out of Solid (Victorian terraces incorporating a mixture of uses, 150 Dwellings / Ha), b) mid 20th century: Buildings as Pavilions (1960's system-built housing floating in barren wastes, 37 Dwellings / Ha), and c) at the beginning of the 21st century: A Denser Urban Fabric (A proposed return to terraced streets with a mix of uses, 75–87 Dwellings / Ha); Hulme Regeneration Ltd; Manchester (June 1994).

[2] Ellis (1978); Rowe and Koetter (1978).

Another related theme in inner city regeneration today is mixed use (as opposed to zoning). Mixed use means mixing living, moving about and working.[3] It takes as its model what we would like to think of as the old, traditional life of cities, that stressed density, multiple use and diversity which is epitomised in historic Clerkenwell. The architectural research has therefore focused on the composition of the urban blocks in each of the thirty housing estates and residential developments, in terms of their mix of uses, built forms, building footprints and faces (*façades*, elevations).

The Public Realm

The specific focus of the paper will consider how the faces of the urban blocks in each of the thirty case studies define and construct the public realm of open space though which people (and traffic) move between buildings. In many parts of the world, housing and neighbourhood developments of the late twentieth century have been shaped by the belief that it is important to create communal open spaces that can provide a physical focus for local communities. This has given rise to a range of modern housing typologies that are based on an ethos of building or creating 'community through design', using concepts such as a 'privacy gradient', the separation of pedestrian and vehicular movement and the notion of 'enclosure', amongst others, to shape the public open space of the development. This in turn gives rise to different degrees of community co-presence and mutual awareness that might vary from 'living in a goldfish bowl'—a state of mutual surveillance where neighbours are very conscious of one another all of the time—to 'living in perpetual night'—where people feel very cut off and isolated from one another in their homes and have no way of relating spatially to other people who live nearby, or even of knowing who their neighbours are.

As Robin Evans so eloquently put it, "If anything is described by an architectural plan, it is the nature of human relationships, since the elements whose trace it (i.e. the plan) records—walls, doors, windows and stairs—are employed first to divide and then selectively to reunite inhabited space".[4] It is these 'traces' in the plan and layout of the sample of housing estates and residential developments that the architects are analysing, in order to understand the mechanism through which public (community, open) and private (dwelling, interior) spaces are first divided

[3] Rowley (1996).
[4] Evans (1997).

from one another by the facades of buildings and then selectively reunited by the windows that overlook public space and the doors that give onto it.

Ways of Seeing

Whilst the architects are finding new ways to visualise these more objective spatial properties of residential areas, the artists attached to the project have independently been developing visual narratives that describe these changes in the urban fabric and their shifting meanings for the ordinary people who inhabit the urban landscape. Helen is particularly drawn to the ‘shifting land use and occupation’ in Clerkenwell. What interests her about the identity of an area is the question that the novelist and biographer Peter Ackroyd posed in his ‘biography’ of London.[5] For Ackroyd, London is a living organism with its own laws of growth and change, so he deliberately refers to his narrative account of 2000 years of London’s growth and change as a biography not a history.

Ackroyd asks, if there is a continuity of life, or experience, is it connected with the actual terrain and topography of the area? Is it too much to suggest that there are certain kinds of activity, or patterns of inheritance, arising from the streets and alleys themselves? This has led Helen to the idea that, in many cases, good intentions in architecture and urban design flounder at street level. This seems to be due to the assumption that a holistic approach to urban planning and design is required. Helen’s idea about the continuity of the city fringe experience, has therefore led her to question the assumptions of architects and planners, and even those of the VivaCity project itself.

Helen sees time as a changing factor that alters the effectiveness of design solutions. This has put her in mind of Godard’s *Deux ou trois choses que je sais d'elle* (Two or Three Things I Know About Her, 1967), and the social failings of the Paris housing complexes. In the film, Godard describes the impact of these good intentions as an anxiety that this architecture provokes. The problematics of improvement and change seem to relate to the dynamic between the city and the fringe. The question is therefore whether one can legislate for this dynamic, and whether legislation itself is a one-way process, something that comes through central policy or that grows up more organically from the grass roots.

Jessica, on the other hand, is intrigued by the clues that are woven into the environment about how to choose, use and experience spaces and

[5] Ackroyd (2000).

places, and how this can be enhanced or restricted by the impact of light, smells, sounds, history, mood, company and time.

Seeing and Going

To this end, both the architectural and the artistic sides to the collaboration have evolved novel ways to represent and portray the precise nature of the boundary or interface between public (street or open space) and private (interior of the dwelling) space, as it is this boundary above all, that has been manipulated in traditional, Modern and Post-Modern housing typologies.

The methodology used by the architects is based on a detailed analysis of the built forms and spatial layouts of housing estates and residential areas, combined with a ‘space syntax’ analysis of the interconnectedness and accessibility of the public realm within each estate, representations that show the extent of visual fields from key public open spaces and an appraisal of the types of physical boundaries surrounding each of these open spaces and the openings that give onto them. The juxtaposition of these different analyses provides insight into the interfaces maintained spatially between residents and non-residents (visitors to the locality, passers-by) and the possibilities that are thereby created for mutual awareness and encounter. The architects are particularly interested in the relationship between ‘seeing and going’ that is constructed by the way that the windows of individual dwellings overlook or survey public space and people’s front / back doors open directly onto or constitute public space.

In his seminal text *Discipline and Punish*, Michel Foucault[6] highlighted the transformative, disciplinary potential of surveillance (seeing), explaining the power structures that are inherent in surveillance techniques. Surveillance is a key theme in the literature on panopticism, alongside that of the ‘gaze’. In his description of a plague city, Foucault even suggests the mechanism by which space and surveillance can come together to express power relations through a process of spatial separation, segregation and partitioning that physically immobilizes people to construct both a counter-city and the perfect society.

Going (the ability to directly access one space from another, sometimes referred to in space syntax theory as constitutedness) embodies the opposite principles of integration, movement, co-presence and encounter. It has received less attention than surveillance, but it is recognised as an important component of urban safety as well as providing opportunities

---

[6] Foucault (1977).

for socialisation and community formation. The purpose of capturing and representing the potentiality of space for seeing and going is shared with the artists in residence, who have been working towards alternative representations of the same phenomena, particularly in respect of the concept of natural surveillance.

Natural Surveillance

Natural surveillance is a term used in Crime Prevention Through Environmental Design (CPTED)[7] and 'Defensible Space'[8] models for crime prevention. These models rely on the ability to influence offender decisions preceding criminal acts. Research into criminal behaviour has shown that the decision to offend or not to offend is influenced more by cues to the perceived risk of being caught than it is by cues about potential reward or ease of entry. Any architectural design that enhances the chance that a potential offender will be, or might be, seen is a form of natural surveillance. Often, it is not just the fact that the offender will be seen that matters. It is that offenders 'think' they will be seen that can help deter the opportunity for crime.

Natural surveillance therefore limits the opportunity for crime by taking steps to increase the perception that people can be seen. It occurs through the design and placement of physical features, activities and people in such a way as to maximize visibility and foster positive social interaction. Potential offenders feel increased scrutiny and limitations on their escape routes. The idea was first formulated by Jane Jacobs in *The Death and Life of Great American Cities*[9] based on her work in New York's Greenwich Village, itself an organic (naturally occurring, unplanned) urban village like Clerkenwell.

Natural surveillance is naturally occurring and it involves 'going' or space occupancy as well as 'seeing' or space observation. As people are moving around an area, they will be able to observe what is going on around them, provided the area is open and well lit. Supporting a diversity of uses within a public space is believed to be a highly effective natural surveillance strategy. Other ways to promote natural surveillance include low landscaping, good levels of street lighting at night, street designs that encourage pedestrian use and designing out any places where criminals can lurk or hide. Included in the design are features that maximize

---

[7] Tilley *et al.* (2005).
[8] Newman (1973).
[9] Jacobs (1961).

visibility of people, parking areas and building entrances: doors and windows that look out on to streets and parking areas, see-through barriers (like glass brick walls, ironwork railings, timber fences), pedestrian-friendly streets and pavements and front doors / porches that give onto public space, all features that have been studied and quantified by the architects in respect of each of the thirty housing projects that make up the project's urban residential database.

Community by Design

Natural surveillance is a relatively recent spatial phenomenon, but for many years, indeed, ever since Clarence Perry[10] articulated the 'neighbourhood unit formula' in the 1920s, architects and planners have recognised that to create 'livable' neighbourhoods it is necessary to ensure that residential districts are designed in such a way as to include a diversity of potential residents, and in ways that facilitate community formation. This means taking account of the requirements of a diverse population: young and old, wealthy and poor, able-bodied and disabled and people of different ethnic backgrounds. To this end, it is believed that a well-designed, balanced neighbourhood should contain a diversity of housing types and tenures, to enable people from a wide range of socio-economic levels, age groups and ethnicities to live together within its boundaries and form a cohesive local community. Some suggest that the size of the neighbourhood should be determined by the population that will support a local primary school.

To discourage through-traffic and to protect the interior from penetration by non-residents, the neighbourhood itself should be bounded by arterial streets. Access points should be chicaned and as far as 'outsiders' are concerned, the interior of the neighbourhood should appear labyrinthine. Local streets, pedestrian paths and cycle routes, on the other hand, should contribute to a system of fully-connected and interesting routes that connect all potential destinations. Their design should encourage pedestrian and bicycle use by being small and spatially defined by buildings, trees and lighting. Likewise, it is believed to be important that as well as housing, the neighbourhood should create a complete community with a full complement of local shops, workplaces, schools, parks and community facilities that are necessary if the everyday needs of its residents are to be met. Its size should be such that housing, jobs,

[10] Perry (1929).

everyday needs and other amenities are within easy walking distance of one another and of public transport.

Grouping housing around a central open space is considered to be an elementary spatial gesture that encourages community formation. To support community development, in addition to a well-defined edge the neighbourhood should therefore have a clear central focus that combines local commercial, civic, cultural, spiritual and recreational uses. Some designers apply this principle recursively, with small clusters of houses grouped around a shared focal space at the base of a hierarchy that then groups housing clusters around a larger, shared space that represents a residential 'quartier', and 'quartiers' around an even larger neighbourhood space that corresponds to the whole community. Uses that integrate people from outside the neighbourhood, on the other hand, should be located at the edge.

It follows that a well designed community should also contain an ample supply of specialized open spaces in the form of squares, greens and parks whose frequent use is encouraged through spatial placement and design. Public spaces should be designed to encourage the surveillance and presence of people at all times of the day and night. Finally, these principles should be expressed in a distinctive architectural style that reinforces the uniqueness of the neighbourhood and encourages local residents to identify with their spatial community. Visualisation techniques have been developed by the architects to record and measure all these properties, in order to compare the morphology of the thirty estates in the database. The output will then be compared to resident-generated assessments of the 'liveability' of each residential area.

Alternative Approaches to Natural Surveillance and Community

Intrigued by the concept of natural surveillance, artist Helen has been making a special study of windows, as they reflect notions of public and private space. She has begun to shoot images of window frontages for a video montage on the theme of public and private space, which looks specifically at the construction of boundaries and borders between our homes and the wider community, see figure 5.

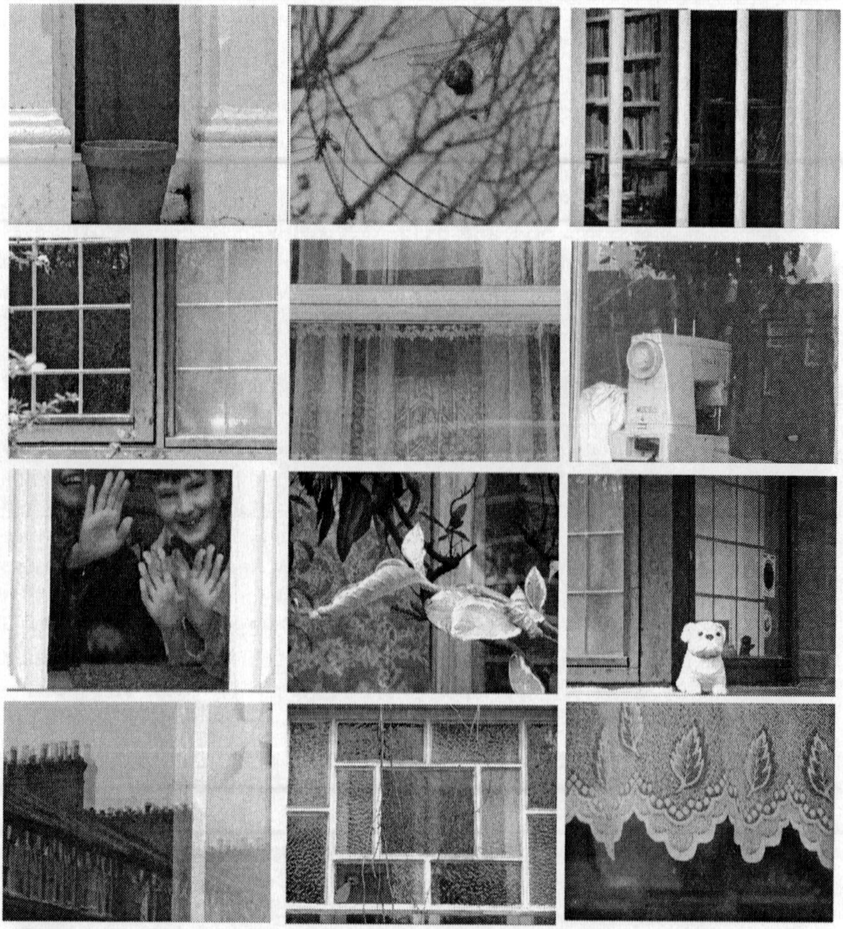

Figure 5: Definition of the urban public/ private boundary by windows, a study of Clerkenwell, London.
Source: Original artwork by Helen Bendon

Another of Helen's pieces is about the continuity of experience in the city fringe. Through a series of visually linked small narrative moments, a sense of continuity, movement and progress is formed as the gestures and movements of women are captured in a fragmented narrative. These women going about their activities are of the fringes; there is poverty, crime, prostitution, sadness, stillness but ultimately a sense of survival. The series of narrative works is connected together through the spaces of

Clerkenwell, creating an overall atmosphere of the area that is about the seedy underbelly of city life, see figure 6.

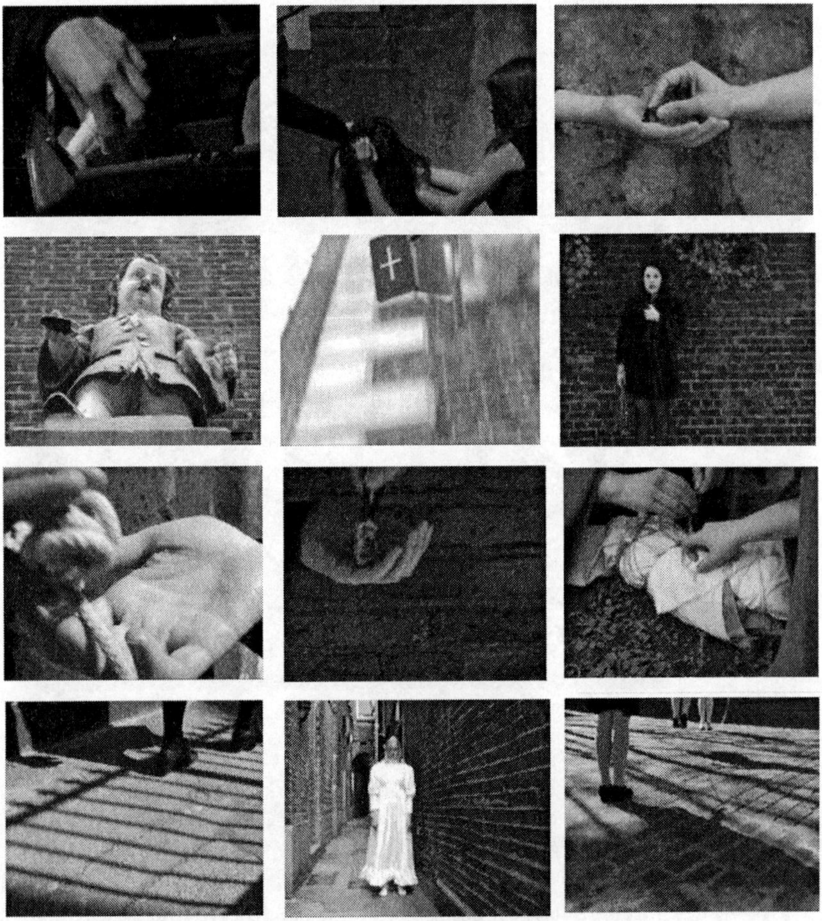

Figure 6: A portrayal of women's lives in Clerkenwell.
Source: Original artwork by Helen Bendon

A third strand in Helen's work explores the experience of living in a community like Clerkenwell, but her interviews have led her to think about 'entrapment' rather than 'community by design'. This is based on the experience of many of the residents resisting a move or being unable to move, living lives that are in sharp contrast to the notion of progress and development in a 'successful mixed-use urban area'. To express this sense of personal stasis in the face of rapid urban change, Helen has evolved a video narrative technique where there is little or no movement, due to the actors being excluded from 'progress'.

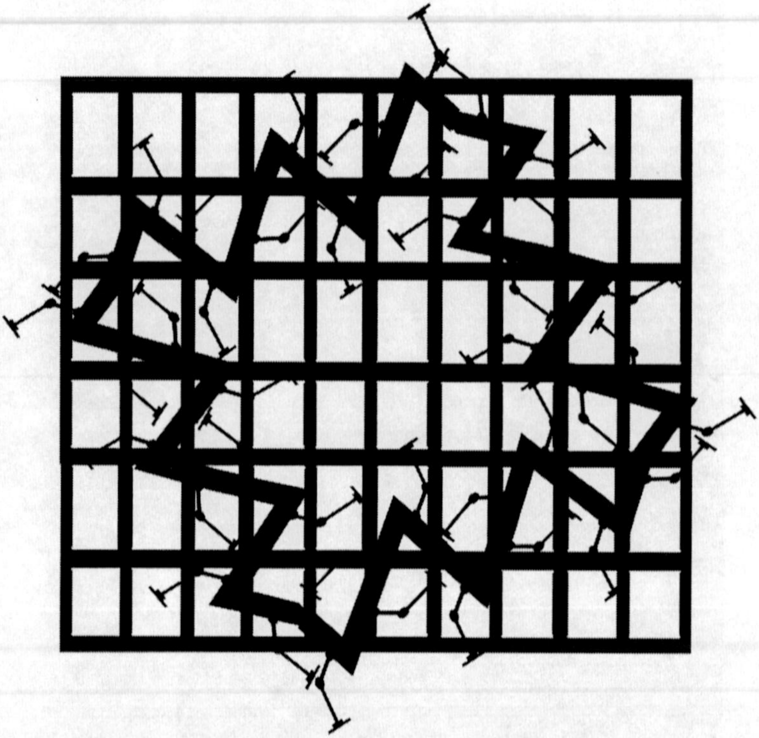

Figure 7: CCTV surveillance in Clerkenwell, London.
Source: Original artwork by Jessica Thom

Jessica, on the other hand, has been mining out the content of the database of nearly three thousand photographs that have been collected as part of the VivaCity project. She has analysed the content of each image, counting how many times various phenomena—such as windows, CCTV cameras, people, cars, and so on – feature, and she then has used images extracted from the photographs to re-present this information to the viewer in a series of 'portraits' of Clerkenwell that are a fusion or summary of many different people's experiences. The research has taken over sixty thousand images of windows, ten times as many as of any other features of the built environment. It includes nearly two thousand images of people and one hundred and eighty-four of CCTV cameras that are lurking in the background of the photographic record, observing and recording the researchers on film as they observe and record the environment photographically—a blurring of surveillance with something like sousveillance. *Souveillance* (inverse surveillance) has been defined as the surveillance of those in authority by those who are normally watched, a covert reversal of normality which may be enabled by the situating of miniaturised surveillance technology like wearable cameras on the clothing of the people who are being observed[11] (see figure 7). In this case, the CCTV cameras watched the architects as they recorded the residential areas, but Jessica has observed their presence and has subversively re-contextualised and re-presented their ubiquitous gaze.

Database of Residential Districts

With these ideas in mind, attention can now turn to the database of residential districts that the architects have constructed in order to carry out spatial analysis. Currently, it comprises twenty estates from Clerkenwell and five each from Manchester and Sheffield. The oldest examples of housing are conserved areas of traditional street system lined with terraced houses, dating back to the 1820s and 1830s. Three are from London and one is from Manchester. The most recent comprise innovative, mixed use urban regeneration projects from London, Sheffield and Manchester.

Particularly in the inner city, these projects are believed to be a vital ingredient in creating successful, efficient, sustainable and vibrant communities. The spaces between the buildings, the interrelationship between the different uses and the creation of a sense of place are all thought to be key to designing successful urban regeneration projects, and in this respect the projects that have been selected for analysis represent exemplars of a new approach to housing in a mixed use context that is being implemented in the heart of just about every major city in the UK.

Arguably, this housing database is another response to Ackroyd's[12] concept of 'the city as body', in which the byways of the city resemble thin veins and its parks (and green spaces) are like lungs. Ackroyd goes on to state that the whole city is experienced only as a wilderness of alleys and passages, courts and thoroughfares, in which even the most experienced citizen may lose the way, rather like inhabiting a body that is

[11] Mann *et al*. (1998).
[12] Ackroyd (2000).

unfathomable or out of control. The research aim is to unveil the global form and lineaments of these diverse housing areas through careful analysis.

Figure and Ground

The research methodology adopted by the architects relies heavily on GIS-type computer analysis and representations that permit the quantification of built form variables, a set of techniques known as collectively as 'space syntax'. This form of analysis is about pattern recognition—something that the human eye and brain is very good at—and so numerical measures have been translated into colours of the spectrum, so that it is immediately possible to appreciate the spatial distribution of numerical values and the emerging patterns that these make. Because these spatial properties are measurable, they can be regarded as research findings, unlike the work of the artists, which inevitably must remain more speculative as they respond to the science by presenting their initial thoughts and ideas.

Many 'syntactic' measures have been shown by previous research conducted over the last 30 years to reflect observed patterns of co-presence in and (particularly pedestrian but also vehicular) movement through space—in other words, how busy or quiet it is likely to be, so the analysis can directly address and predict how the estates and residential developments are likely to perform in terms of concepts like natural surveillance and human co-presence in open spaces designed for community use.[13]

The foundation image in 'space syntax' is a figure/ground ratio for each of the thirty housing schemes, produced for each residential area in an identical way to the composite image of Hulme, shown earlier in figure 4. This measures the total area of the building footprint for each urban block, and compares it with the amount of unbuilt space that is left over. The higher this ratio, the more built-up the block; the lower the ratio, the more open space there is left around the buildings. Given what has been said already about the transformations that have occurred in the urban tissue over the last 100 years, it is to be expected that housing from different periods will have different figure/ground ratios that reflect the different architectural philosophies of the day, and indeed this is the case (see figure 8). The examples in figure 8 are arranged from left to right in the order in which they were built, so that the earliest examples are to the far left and the most recent to the far right. The average for the whole

---

[13] Hillier and Hanson, 1984; Hanson, 1996.

sample is 0.749, but for the traditional (pre1899) urban blocks it is 0.609, the early Modern (1900-1945) and High Modern (1945-1980) periods have rather lower ratios of 0.497 and 0.565 respectively, whilst for the Post-Modern (post 1980) period the figure/ground ratio is 1.261, twice as high as in the traditional urban blocks.

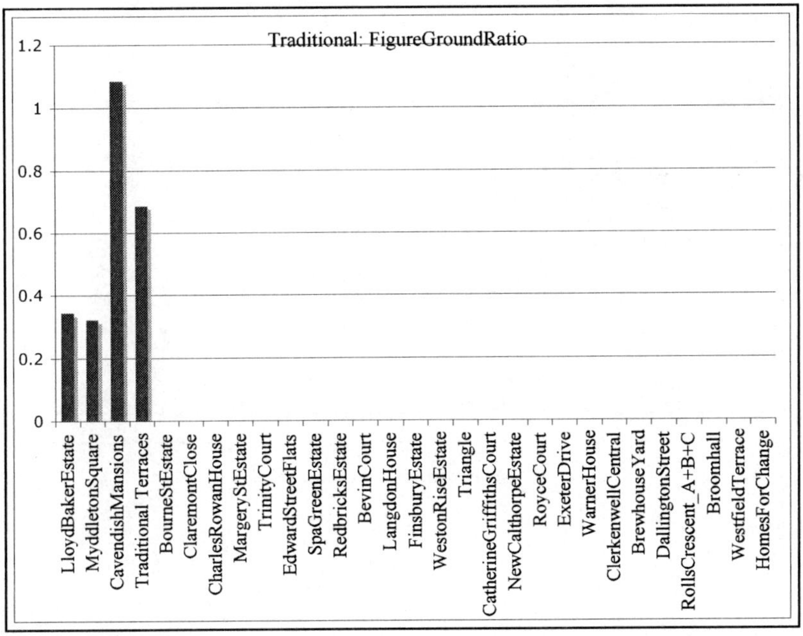

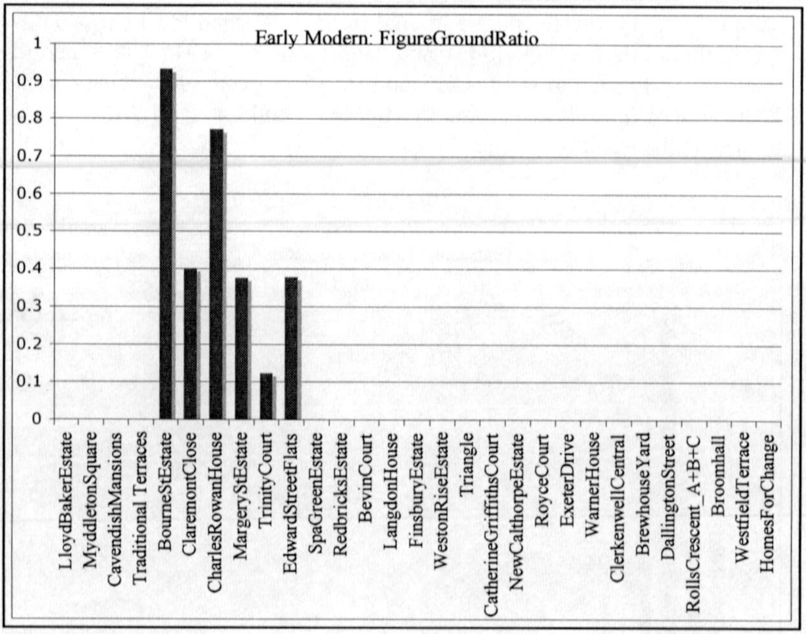
Early Modern: FigureGroundRatio
1
0.9
0.8
0.7
0.6
0.5
0.4
0.3
0.2
0.1
0
LloydBakerEstate
MyddletonSquare
CavendishMansions
Traditional Terraces
BourneStEstate
ClaremontClose
CharlesRowanHouse
MargeryStEstate
TrinityCourt
EdwardStreetFlats
SpaGreenEstate
RedbricksEstate
BevinCourt
LangdonHouse
FinsburyEstate
WestonRiseEstate
Triangle
CatherineGriffithsCourt
NewCalthorpeEstate
RoyceCourt
ExeterDrive
WarnerHouse
ClerkenwellCentral
BrewhouseYard
DallingtonStreet
RollsCrescent_A+B+C
Broomhall
WestfieldTerrace
HomesForChange

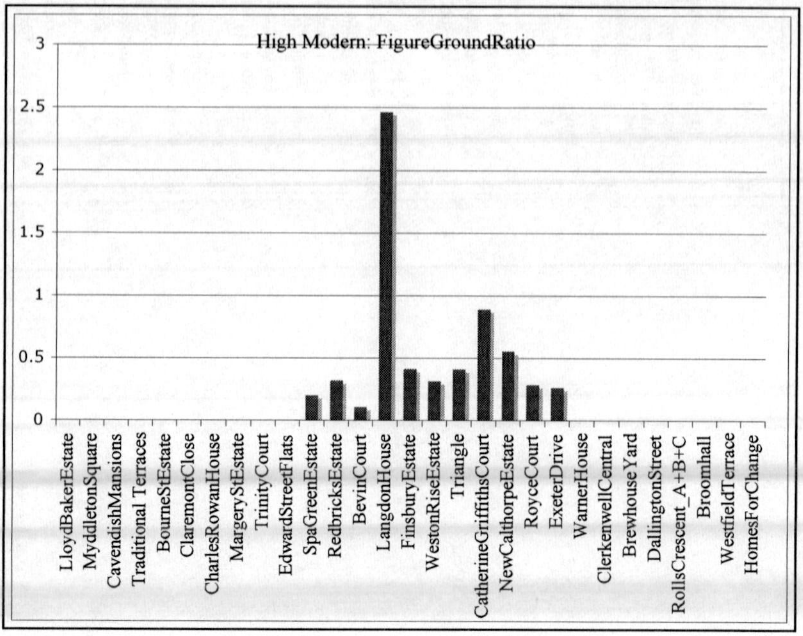
High Modern: FigureGroundRatio
3
2.5
2
1.5
1
0.5
0
LloydBakerEstate
MyddletonSquare
CavendishMansions
Traditional Terraces
BourneStEstate
ClaremontClose
CharlesRowanHouse
MargeryStEstate
TrinityCourt
EdwardStreetFlats
SpaGreenEstate
RedbricksEstate
BevinCourt
LangdonHouse
FinsburyEstate
WestonRiseEstate
Triangle
CatherineGriffithsCourt
NewCalthorpeEstate
RoyceCourt
ExeterDrive
WarnerHouse
ClerkenwellCentral
BrewhouseYard
DallingtonStreet
RollsCrescent_A+B+C
Broomhall
WestfieldTerrace
HomesForChange

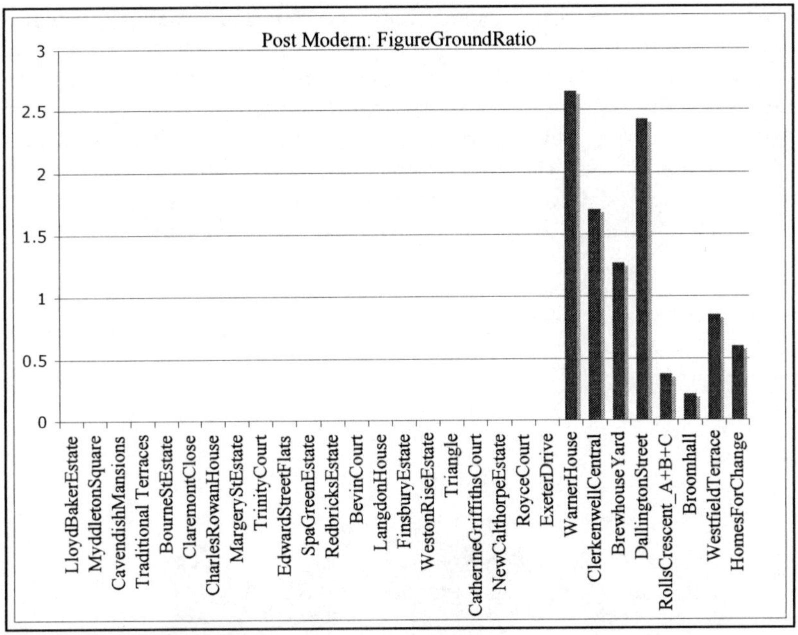

Figure 8: Comparative figure /ground ratios for the housing estates/–developments in the VivaCity urban database, grouped and arranged by the eras to which they belong: a): Traditional (pre 1899), b): Early Modern (1900–1945), c): High Modern (1945–1980), and d): Post Modern (Post 1980).

Looking more closely at the most recent, Post-Modern schemes, they clearly fall into two distinct types: the four most recent examples are low rise new 'street layouts' that are trying through their built form to replicate the urban fabric of the traditional era and the first four Post-Modern examples are the recent innovative mixed use inner-city urban blocks (MUUBs). The two types have very different figure/ground ratios; 0.510 for the streets and 2.012 for the MUUBs. This shows how extreme the site coverage has become in this particular type of Post-Modern development.

Open Spaces

The next representation classifies all the unbuilt open spaces of the urban blocks on which the buildings stand, in order to identify the 'ownership' of the spaces. Figure 9 shows the open space classification of a typical

High Modern housing estate from the housing database. Unclear 'ownership' of spaces in the public domain has been identified with social malaise.[14] The typology differentiates between vehicular roads, pedestrian and cycle paths, areas for car parking, common green areas and hard landscaped spaces, private yards and gardens and areas of restricted access that no one can use. The proportions as well as the areas of each type of space in each estate or residential development have been calculated, in order to eliminate the effects of developments of different metric areas.

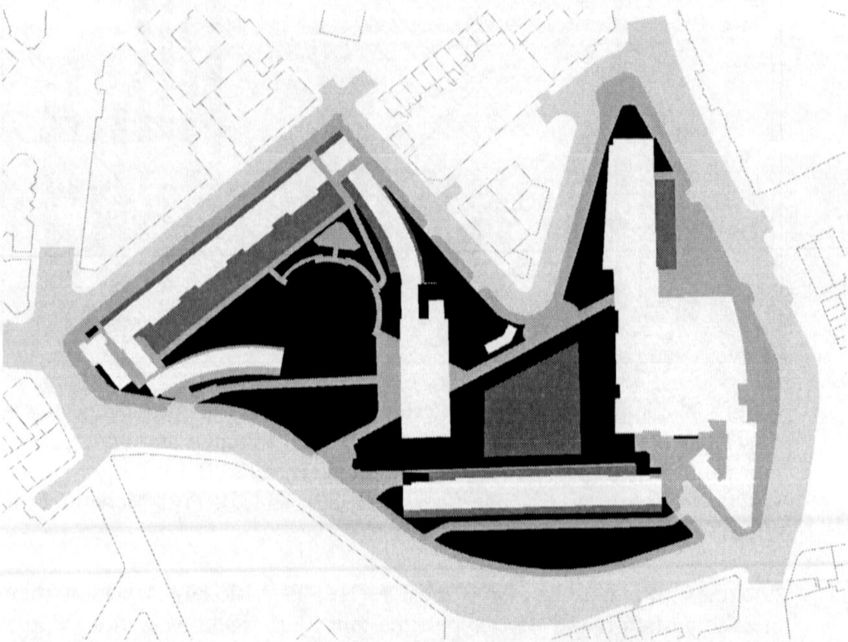

Figure 9: Open space classification of a typical High Modern housing estate from the housing database. The typology differentiates between common green areas (in black and progressing in lighter shades of grey), car parking, private yards and gardens, pedestrian and cycle paths, and vehicular roads (lightest grey). The buildings in this figure are in white.

Unsurprisingly, the most common space type found within the residential or mixed use urban block is paths. All the developments studied have a pedestrian system, but the proportion of open space devoted to public circulation varies from as little 35% in the High Modern era as to as much

[14] Coleman (1990).

as 66% in the Post-Modern mixed use regeneration projects, with an average 44% for all 30 schemes. The traditional era and the Post-Modern era therefore share an interest in creating life though pedestrian movement within hard landscapes and systems of paths.

Following from the preoccupation with creating shared, open spaces for community use, common green spaces are the next most prevalent space type, occupying on average 28% of the unbuilt space in the sample. About a quarter of unbuilt space was dedicated to this use in the two pre WWII eras and it was greatest in High Modernism (39%) but this use has almost disappeared in the Post-Modern period where only 8% of all unbuilt space is dedicated to common green areas. Five schemes, all post WWII have almost no common green space at all. The philosophy of enclosing shared open spaces to promote community use is therefore a characteristic of Modernism.

Private gardens are the next most prevalent space type, occupying on average a quarter of the unbuilt space of the schemes. These are most prevalent in traditional housing layouts, occupying on average 30% of unbuilt space, and are least numerous in the Post-Modern urban regeneration blocks at 7%. However, a third of schemes from across all time periods have few or no private gardens.

Perry[15] was one of the first to observe that the car was one of the most significant factors that has led to changes in contemporary housing morphology and this has been reiterated many times since,[16] but as far as parking is concerned, 40% of schemes make no provision for surface car parking. The average proportion of unbuilt space dedicated to car parking across the sample is 11%, but this masks a profound split between the pre WWII eras, which had 6% of car parking space, and the post war eras, which had 11% and 14% respectively.

The proportion of space that is restricted so that only the residents of the housing can access it, is greater than that devoted to cars, 18% on average, but whilst only two pre WWII schemes had restricted spaces, half of the post WWII have such areas, each accounting for 20% of all the unbuilt space in the High-Modern and Post-Modern periods.

Spatial Integration

The general idea behind 'space syntax' is that the spaces depicted in an architectural plan or layout—in this case the public open spaces within a

---

[15] Perry (1929).

[16] McClusky (1992); Jenks *et al*. (1996); Panerai *et al*. (2004).

housing estate or residential development—can be broken up into components, analyzed as networks of route choices, then represented as maps, graphs and mathematical measures that describe the relative connectivity and integration of those spaces. Adapting Ackroyd's metaphor of urban space as labyrinth, space syntax can be thought of as a wayfinding technique.

Integration is one such mathematical measure that calculates the average depth of each space to all other spaces in the layout. The spaces of a system can be ranked from the most integrated to the most segregated and coloured up from black (most integrated, busy hot spots) to palest grey (most segregated, isolated cold spots). Using UCL Depthmap, the architects have performed a visibility integration analysis of each estate or residential development to show its pattern of integration. Depthmap takes input in the form of the layout of the estate or development and is able to construct a distribution map of 'visually integrated' (black) and 'visually segregated' (pale grey) locations within it (see figure 10).

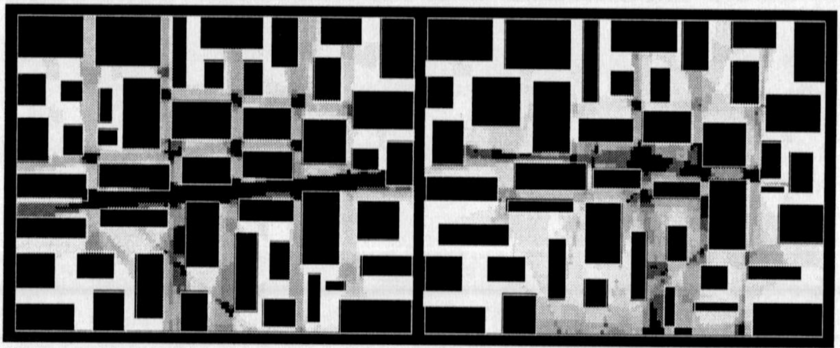

Figure 10: Distribution of spatial integration in two hypothetical urban layouts with the same number of blocks and topological relationships, but with different geometrical relationships that maximise and minimise integration respectively.

In the urban fabric of traditional cities, time and time again integration has been shown to be a good predictor of the actual amount of pedestrian flow passing through a space, and so it is should also be a good indicator of the potential that the public realm of a neighbourhood has to generate co-presence among neighbours and between local people and passers-by, and these empirical results can also be replicated in agent-based simulations in virtual environments. Figure 11 shows the emergent density of trips selected by many hundreds of randomly-distributed avatars in a computer generated simulation of pedestrian movement around the two layouts previously shown in figure 10. Note that the relative density of route choice, graded from black (most popular route segments selected) to pale grey (least popular) corresponds closely to the pattern of spatial integration.

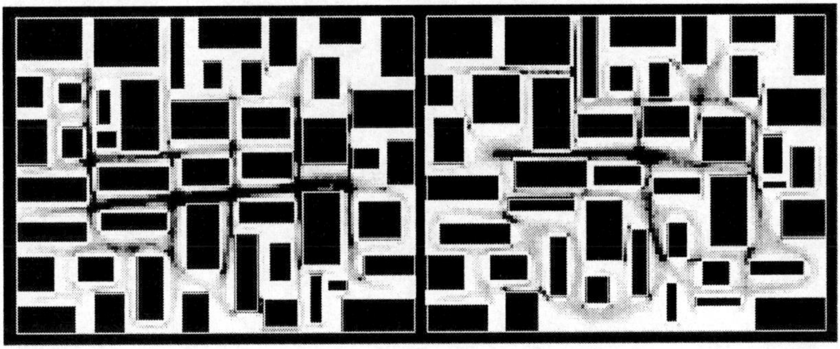

Figure 11: Distribution of movement traces of randomly moving avatars in the two hypothetical urban layouts previously illustrated. Note the correspondence between higher (darker grey) movement density and higher spatial integration in each layout.

Successful shared spaces should be appropriately integrated into the layout to form a natural catchment for the relevant level in the community hierarchy, but all too often because integration is difficult to conceptualise intuitively, architects and urban designers produce layouts in which key community spaces are over segregated or over integrated for their function. The analysis is intended to unveil this relationship between spatial design and social purpose.

Integration analysis revealed that half the schemes in the database were visually integrated communities that had a large black, therefore integrated, focus at or close to the geometric centre of the urban block (see figure 12). These were mainly early or High-Modern housing estates. However, a third of the schemes in the database, had no clear focus at all but exhibited a peripheral pattern of spatial integration that was confined to the surrounding streets (see figure 13). Most of these were either traditional nineteenth century urban neighbourhoods or late twentieth and early twenty-first century schemes. A final group of five schemes had no clear overall focus for integration, but showed a diffused network of internal circulation with some integrated (black, hot) spots and other more segregated (cool, pale grey) areas within an overall village-like layout (see figure 14)

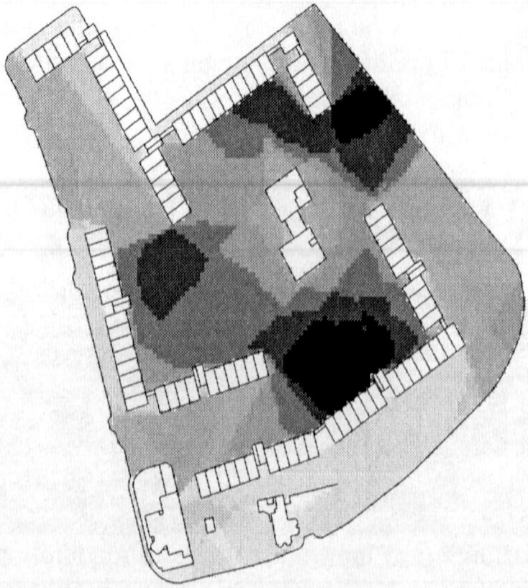

Figure 12: Centralised pattern of spatial integration in a typical visually integrated community

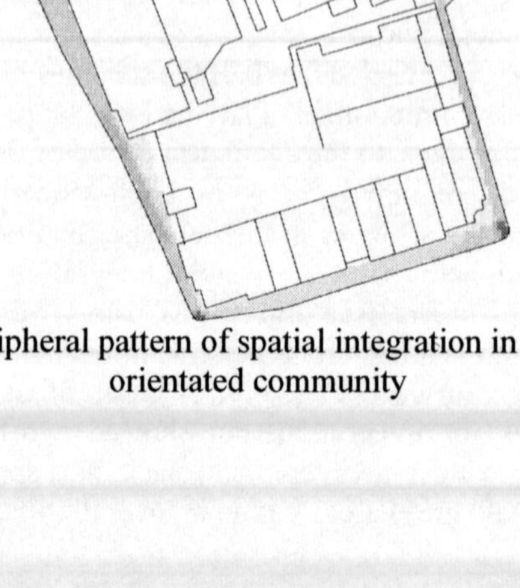

Figure 13: Peripheral pattern of spatial integration in a typical street orientated community

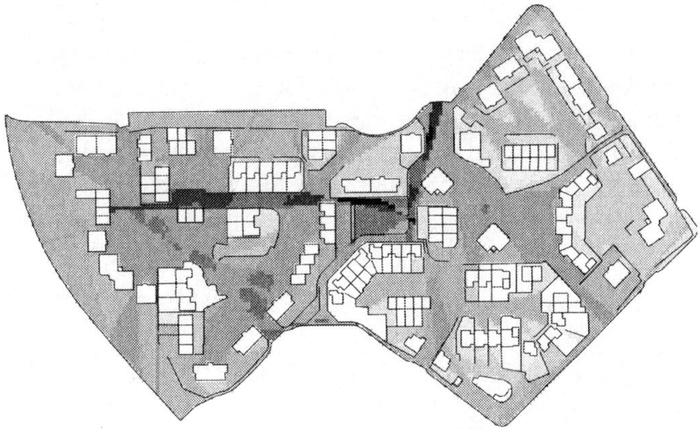

Figure 14: Dispersed pattern of spatial integration in a typical urban village

Surveillance

The final set of measures attempts to grapple with the issue of natural surveillance by recording exactly how each building *façade* is constructed, in terms of its height, transparency or opacity, and permeability with respect to doors and windows. This indicates the potential each *façade* has to allow occupants to overlook the adjacent open spaces out of doors, or to access them directly. Primary (building) boundaries are differentiated according to whether they are comprised of active shop frontages and doors and windows that people can cross and that link the public exterior sphere to the private interior domain of people's dwellings, windows and upper level visibility only so that the public / private boundary thus constructed is one of seeing or surveillance but not of permeability, or blank walls that permit no relationship whatsoever between the public and private worlds on either side of the boundary. Secondary (open space) boundaries or barriers are differentiated according to whether they are a high opaque fence, a high see through fence, a low fence or a very low fence that one could step over.

This analysis reveals that the dominant surveillance characteristic of High-Modernism is that of overlooking, in that large, open, unfenced green landscaped areas are overlooked by the upper level windows of the surrounding blocks of flats. The defining characteristic of Post-Modern street developments is that the interface between the dwellings and the street comprises mainly secondary boundaries made up of the fences and gateways leading to the entrances of houses and apartments, whilst the dominant characteristic of the MUUB schemes is that the interface between buildings and street space at ground level comprises mainly active frontages (doors and windows) for non-residential uses such as shops and offices. These ideas directly intersect with Helen's (mainly Modernist) study of windows, and how they construct both a boundary and an opening that creates an interface between private and public space, and with Jessica's (mainly Post-Modern) portrait of Clerkenwell as a nexus of windows and CCTV cameras, juxtaposing natural and artificial surveillance.

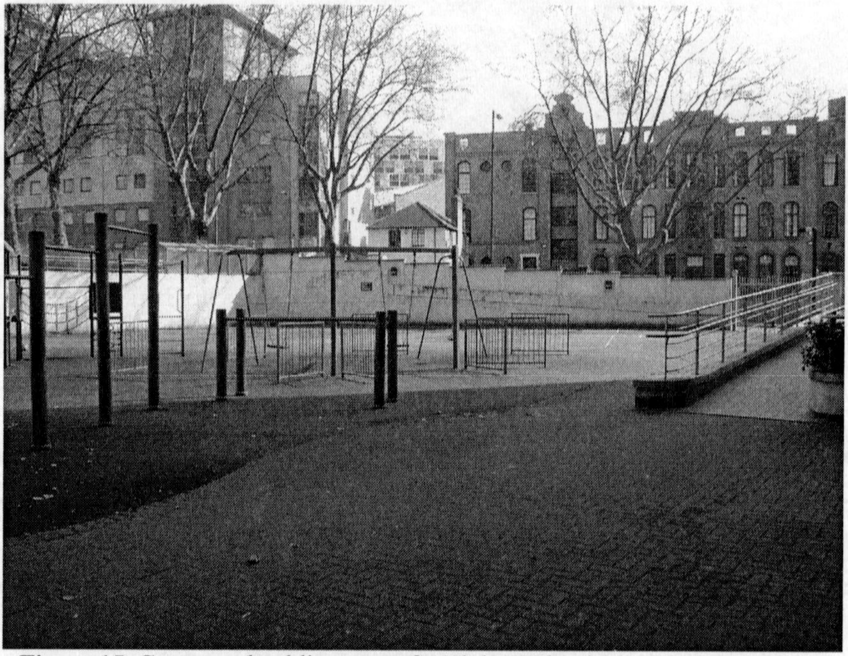

Figure 15: Communal public space of a High-Modernist housing estate in London

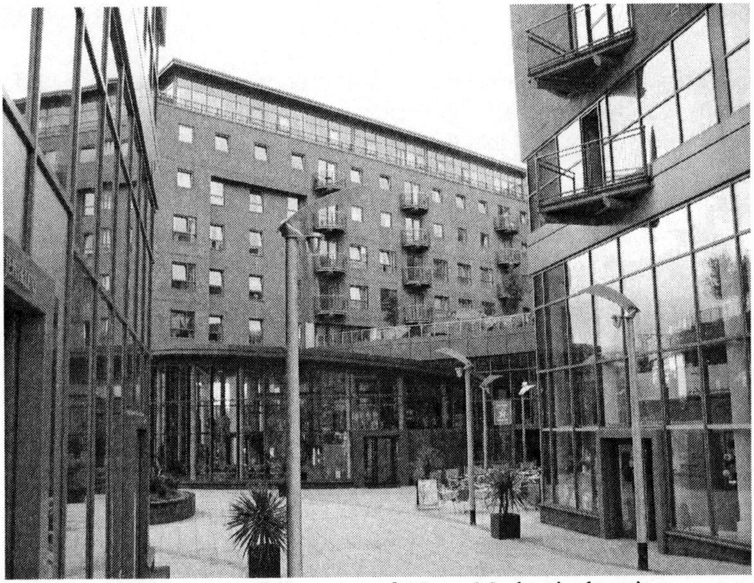

Figure 16: Communal public space of a Post-Modernist housing estate to emulate the squares of the traditional urban fabric

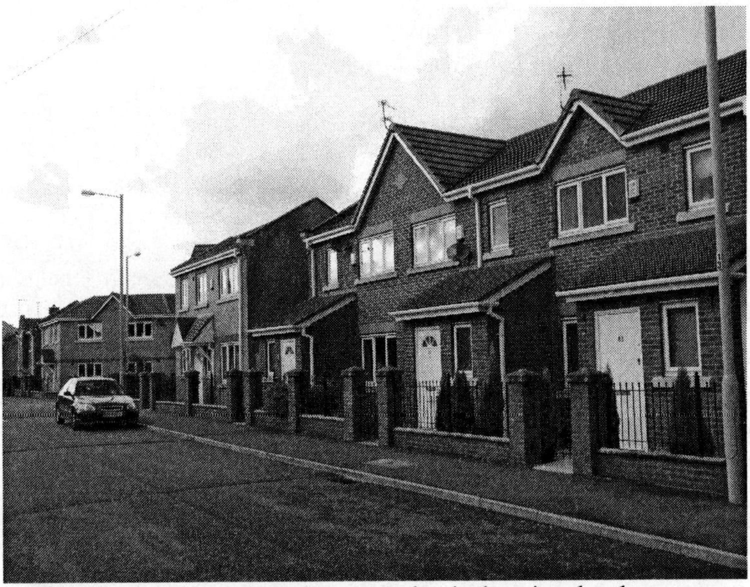

Figure 17: Street-oriented Post-Modernist housing developments

Conclusion

The architects and the artists have only been working together for a few months, but the relationship is already beginning to bear fruit. The artists have already discovered that the initial stages of becoming involved with a research project have exposed them to a great range of material that raised questions about how to integrate the issues of the physical city with the everyday experience of its users – in essence it has challenged them to develop ways to express the notion of a 'social city'. The architects have been encouraged to humanise their objective spatial representations and measures, by developing visually intuitive schema that capture what it is like to journey through the spaces and places of the residential urban environment.

Perhaps what most distinguishes architecture from painting, photography and video is its spatiality. The history of domestic architecture is primarily a history of shaping spaces, both public and private, by enclosing spaces, defining volumes and constructing *façades* with doors and windows that permit different modes of access, egress and overlooking to serve different social purposes. Syntactic analysis has shown that after a period of post-war architecture dominated by High-Modernism, today's urban fabric is, if anything becoming tighter and denser, emphasising proximity, paths and physical connections over the open and amorphous green spaces that were supposed to encourage community formation but which in practice produced little more than a symbolic 'community of the eye'.

This does not mean that community is less important to contemporary society, but rather that it is being interpreted in new residential typologies. In this respect, the dwelling place of home and neighbourhood are arguably more, not less, important to many people's lives and this appears to be signalled by a resurgence of interest in new forms of well-integrated, 'street' or 'square' typologies for residential areas, as opposed to enclosed, open-centred housing 'estates'. The street layouts develop mixed uses around the perimeter of the urban block so that there is no real focus for community life other than the natural, unforced co-presence of residents and passers-by that occurs there. The alternative MUUB produces one or more hard landscaped *piazzas* in the interior of the urban block that facilitate interaction between local residents and visitors to the city centre. Both these typologies rely on creating a rich mix of residential and non-residential uses locally, and neither relies on a hierarchy of open green spaces as a means of facilitating community formation.

In this respect, an important shared theme has emerged between architecture and art, which has to do with precisely how a physical and spatial interface is constructed and maintained between the private spaces inside people's homes and the collective public domain of the city that is equally the domain of strangers as it is of inhabitants. From Webber's[17] formulation of the idea of a 'non-place urban realm' that results in 'community without propinquity' to Mitchell's[18] 'city of bits' and 'e-topia', contemporary urban theorists have suggested that the specific properties of place have little relevance for modern city dwellers as the world becomes governed less and less by physical boundaries and more and more by abstract networks of technologically-enabled connections. However, this collaboration is beginning to suggest otherwise.

References

Ackroyd, P. (2000) *London: the biography* London: Chato and Windus.

Coleman, A. (1900) *Utopia on Trial, Revised Edition* London: Hilary Shipman.

Ellis, W. (1978) 'The spatial structure of streets' In S. Anderson (ed.) *On Streets* Cambridge MIT Press.

Evans, R. (1997) 'Figures Doors and Passages' In *Translations from Drawing to Building and Other Essays* London: Architectural Association.

Foucault, M. (1977) *Discipline and Punish: The Birth of the Prison* Harmondsworth: Penguin Books.

Hillier, B. and Hanson, J. (1984) *The Social Logic of Space* Cambridge: Cambridge University Press.

Hillier, B. (1996) *Space is the Machine* Cambridge: Cambridge University Press.

Hulme Regeneration Ltd. (1994). *A guide to development, Hulme,* Manchester.

Jacobs, J. (1961*) The Death and Life of Great American Cities* New York: Random House.

Jenks, M., Burton, E. and Williams, K. (eds.) (1996) *The Compact city: a sustainable urban form?* London: Spon.

---

[17] Webber (1964).

[18] Mitchell (1996, 1999, 2003).

Mann, S., Nolan, J. and Wellman, B. (1998) 'Sousveillance: Inventing and Using Wearable Computing Devices for Data Collection in Surveillance Environments' In *Surveillance & Society* **1**(3): 331–355
McClusky, J. (1992) *Road Form and Townscape* Oxford: Butterworth-Architecture.
Mitchell, W.J. (1996) *City of Bits* Cambridge: MIT Press
—. (1999) *E-topia: Urban Life, Jim—But Not As We Know It* Cambridge: MIT Press.
—. (2003) *Me++ The Cyborg Self and the Networked City* Cambridge: MIT Press.
Newman, O. (1973) *Defensible Space: people and design in the violent city* Oxford: Architectural Press.
Panerai, P., Castex, J., Depaule, J. and Samuels, I. (2004) *Urban Forms: the death and life of the urban block* Oxford: Architectural Press.
Perry C.A. (1929) 'Neighbourhood and Community Planning' In *Regional Survey of New York and its Environs, Volume VII* New York: New York City (22–140)
Rowe, C. and Koetter, F. (1978) *Collage City* Cambridge: MIT Press.
Rowley, A. (1996) 'Mixed Use development: ambiguous concept, simplistic analysis and wishful thinking?' In *Planning Practice and Research* **11**(1): 87–99.
Tilley, N. (ed.) (2005) *Handbook of Crime Prevention and Community Safety* Devon: Willan Publishing, Cullompton.
Webber, Melvin M. (1964) 'The urban place and the non place urban realm'. In *Explorations into Urban Structure*, University of Pennsylvania Press, with Maps, Diagrs., Charts, Tables.

CHAPTER FIVE

SEMIOTICS OF SPACES: CITY AND LANDART

ELISEBHA FABIENNE PLATZER

Introduction: Cities

5000 years of documented age and innumerable names: Ur, Uruk, Khorsabad, Assur, Nineveh, Babylon, Karphi, Knossos, Lato, Thera, Syphnos. Cities are special spaces. Their structure is in continuous transformation, renewal and redefinition. From the semiotic point of view urban space has to be analyzed as a discourse; a significant practice. The space is defined as a readable and rewritable text. An ontological etymology defines the city as textum in the sense of texture, and testis, testimony. A weaved tissue of streets, buildings, advertising, churches, people, … witness of historical, anthropological, artistic stratification of signs.

City and stratification

Signs witness the dyachronical stratification. City's genesis is revealed by paleolitic wall paintings and by the neolitic agricultural villages. The introduction of written language, astronomy, mathematics, the cooper working, the use of sailing boats and the loom. The agora, the sanctuary, the gymnasium and the theatre structure the ancient Greek city. The Hellenistic urban vision is constituted by geometric rigour, wide streets, exceptional height, vast public spaces. The Hellenistic culture projects his canon into Roman urbanism: visible symbols of the Empire, aqueducts, viaducts, triumphal arcs, temples, amphitheatres, …With the decline of the Roman Empire the unitarian political and territorial system is split in many little medieval towns with their castles, cathedrals, market places. Assisi, Chioggia, Ferrara, Florence, Gubbio, Montagnana, Orvieto, San Gimignano, Peruse, Siena, Urbino, Venice, Verona. Avignon, Carcassonne, Chartres, Poitiers, Reims, Troyes. Bamberg, Heidelberg, Lübeck, Nürnberg, Rothenburg, Würzburg. Many cities during

Renaissance and Baroque can be considered as mature developments of pre-existing medieval towns. Renaissance introduces paradigmatic changes: figurative programmes concerned also about demographic, commercial, productive questions. Maritime commerce and exchanges with the New World become the basis of the new organization of power from local to national.

City and encyclopaedia

Reading a city, as Ugo Volli tells, means understanding its explicit or implicit metalanguage.[1] Streets and districts have names, some buildings are identified by their function (the hospital, the parliament, …), circulation is subject to specific rules (circulation on the left or on the right side), street signs determine order and access to some areas; but certain zones produce their own unwritten laws of cohabitation, the window of a school has the same function as a window of the parliament; some elements are much more significant and only partially functional. The citizen possesses a semiotic urban competence, an encyclopaedia of reference which enable the decodification of signs. Each situation requires the inference of specific frames. Also the notion of encyclopaedia reveals its diacronical facets: from Aristote to Plinius (Naturalis Historia), Vitruv, Porfirius (porfirian tree) Saint Augustine (De doctrine Christiana), Isidorus from Siviglia (De natura rerum, Etymologiarum sive Originum libri XX) Bartolomeo Anglican (Fisiologus), Lullus (Arbor Scientiae, Ars magna), Roger Bacon, Savigny, Ramo, Jean Heinrik Olmsted, Kircher, Schott, Wilkins, Leibniz, Denis Diderot, D'Alembert, Lakoff, Fillmore. The morphological knowledge makes it possible to orientate movements, our material, esthetical and symbolic uses of the city.

So in fact the microcosmos city is a controlled agglomerate of a great deal of differences. The presence of multiple topical elements and their signification and interpretation constitute the conflictual urban text. The signs of permanent architectonical structures (castles, city walls, …) concur with the possibility of movement and expression and with synesthetic signs (sounds, fragrances) of different cultures. The historical stratification demonstrates this war of codes and semantic (how the city has to be, which urban models are considered to be followed, which sense is given to certain forms). The historical flow of artistic epochs reflects the interactions, conflicts and relations between classes, functions, form of

[1] The semiotic analysis of city as textum an its theoretical consequences can be found in: Volli (2005: 5-20).

governments. The renaissance city of merchant is in conflict with the medieval city of nobility. A new occupying power imposes his sovereignty with his walls, castles, and spaces for religious rituals. Skyscrapers fight for their primacy.

Sketching semiotics' gaze on land-art

Landart reacts to the continuous hegemonical architectural and historical tensions of the urban context and decides to choose alternatives. An intense reflection on the theory of forms is the result. Nature becomes the site of expression. Carl André opposes the notion of verticality applied on sculptures preferring round forms. Robert Smithson composes spirals. James Turrell experiments with light. Morris makes art researching archetypes in Egyptian and mycenean cultures. Long decides to cross spaces. The practice of walking becomes constitutive for his work of art.

Art deautomatizes language. Art eliminates striations to create an 'espace lisse', like it is called by Deleuze and Guattari, in order to rewrite on the new space by inventing 'Verfremdungseffekte'. The negotiation of new spaces.

Umberto Eco applies for the first time in 1962 the notion of *opera aperta*, open work, applied on visual, literary and musical work of arts. This concept is a model and a model implicates a specific methodological attitude. In the Lévi-Strauss' sense of the term: we only refer to a model in the degree it can be transformed and handled. A model as a common form for different phenomena, a handicraft tool. Eco proposed with its notion of open work a relation between work and user; a dialectical instant between the structure of the work as a fixed system of relations and the answer of the user as the insertion and repetition of the same system.[2] Roland Barthes tells about this kind of relation analyzing, in an interview for the magazine Tel Quel, Bertold Brecht's theatre:

> (…) Au moment même où il liait ce théâtre de la signification à une pensée politique, Brecht si l'on peut dire, affirmait le sens mais ne le remplissait pas. Certes, son théâtre est idéologique, plus franchement que beaucoup d'autres: il prend parti sur la nature, le travail, le racisme, le fascisme, l'histoire, la guerre, l'aliénation; cependant c'est un théâtre de la conscience, non de l'action, du problème, non de la réponse: comme tout langage littéraire, il sert à formuler, non à faire; toutes les pièces de Brecht se terminent implicitement par un Cherchez l'issue adressé au spectateur au nom de ce déchiffrement auquel la matérialité du spectateur doit le

[2] Eco (1962: 20).

> conduire…le rôle du système n'est pas ici de transmettre un message positif (ce n'est pas un théâtre de signifiés), mais de faire comprendre que le monde est un objet qui doit être déchiffré (c'est un théâtre de signifiants).[3]

So the semiotic notion of open work focuses on the *Kunstwollen* (Riegel) which is defined by Erwin Panofsky as the conclusive and final sense, which occurs in different artistic phenomenona, regardless whether the author's psychological attitudes and decisions are taken consciously or not.[4]

The stress is on the way the question is posed and on its solution. To formulate instead of acting, to decodificate signifiants instead of transmitting *signifiés*. But the word *Kunstwollen* implicates also the potential artistic expression: *wollen* (to want) anticipates the stage of *sein* (to be). Connected with potential is the incipit-gesture, the mythical gesture of foundation of an earthwork. Each foundation implicates the drawing of a line, the definition of topical limits like interiority/exteriority, inclusion/exclusion, memory/amnesia, difference/identity; and temporal limits (from here on the beginning/end is signed). The artistic gesture of foundation reveals the underlying striation, the utopian ground, which the potential is set free.

Mapping Gibellina's architectural system

Subterranean and Submarine visible and invisible lines (perceptions, practices) unify countries and abolish distances. Johan Wolfgang Goethe, on the road from Alcamo to Sciacca in Sicily on 21st April 1787, noted:

> At midnight I had a strange experience. We were dead tired and had thrown ourselves on our beds in hostellery which was anything but elegant. At midnight I woke up and saw over my head a star so beautiful that I thought I had never seen one like it. Its enchanting light seemed a prophecy of good things to come and my spirit felt utterly refreshed, but soon it disappeared. Leaving me alone in the dark. It was not till daybreak that I discovered what had caused this little miracle. There was a crack in the roof and I had woken up just at the very moment one of the most beautiful stars in the firmament was crossing my private meridian.

It is said that the German poet witnessed this miracle in the place where today Gibellina Nuova stands, the symbolic city of the Belice Valley and

---

[3] *Ibid.*

[4] *Ibid.* 19.

of the reconstruction after earthquake of 1968. Some of the most important Italian and European architects and artists of the 20th century like Pietro Consagra (The Belice Star, sculpture which marks the entrance to Gibellina), Oswald Mathias Unghers, Pierluigi Nicolin, (the city lay-out centred around the Town Hall and the adjacent municipal square), Alberto and Giuseppe Samonà (Town Hall), Sergio Musmeci and Ludovico Quaroni (spherical church), Francesco Venezia (Casa di Lorenzo and the Secret Gardens), Fanco Purini and Laura Thermes (the grouped lay-out of the five squares Sistema delle Piazze, the Chemist's House, the Pirello House), Marcella Aprile, Teresa La Rocca and Roberto Collovà (Case di Stefano), Pietro Consgra (building Meeting, the Città di Thebe sculptures in the municipal square and the gates of the cemetery where the tomb of the artist [he died in 2005] can be found), Arnaldo Pomodoro (various sculptures) contributed to form the new city-museum.

The Municipal Museum counts works of Carla Accardi, Mario Schifano, Mimmo Paladino, Rotella, Joseph Beuys, Renato Guttuso, Alighiero Boetti and others. Near the city the Baglio Case Di Stefano; it is the seat of the Orestiadi Foundation that has presented since 1978 annually high culture dramas and artistic events. Inside the Museo delle Trame del Mediterraneo, an extraordinary display of Mediterranean artefacts, textiles, jewellery, and contemporary art. Gibellina portrays its future in order to forget, its faith in contemporary art as reaction to death. Mumford's utopias of reconstruction. Foucault's concept of *eterotopie*.[5] Intersemiotic translation guided by the Senator Corrao and by the population.

---

[5] Foucault (1984).
[...] una sorta di contro-luoghi, specie di utopie effettivamente realizzate nelle quali i luoghi reali, tutti gli altri luoghi reali che si trovano all'interno della cultura vengono al contempo rappresentati, contestati e sovvertiti; una sorta di luoghi che si trovano al di fuori di ogni luogo, per quanto posano essere effettivamente localizzabili. Questi luoghi, che sono assolutamente altro da tutti i luoghi che li riflettono e di cui parlano, li denominerò, in opposizione, alle utopie, eterotopie.

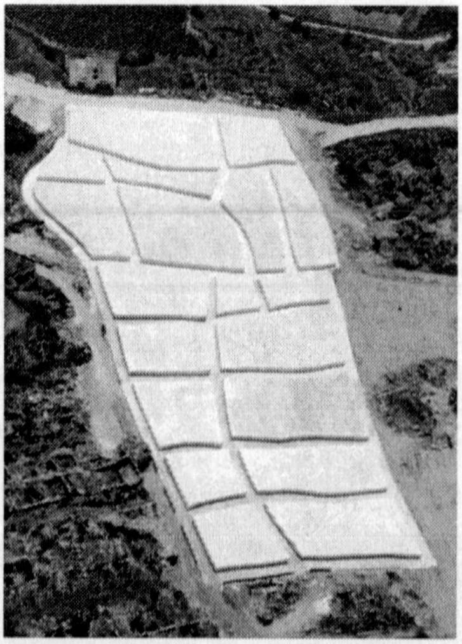

Figure 18: Alberto Burri's Cretto

Some few kilometres away the ancient destroyed city of Gibellina. A site completely reconfigured by Alberto Burri's Cretto. After the earthquake the former city, situated on a hilly landscape, has been totally destroyed. The decision to move away from the seismic zone to a safer location was related to the desire of leaving a significant sign in the sense of sedimentation. A sign able to evoce the experienced sufferings in order to activate the *Trauerarbeit*, a process to digest the sorrow and prepare for the future.[6]

Burri, painter and sculpturer, known for his experience in the concentration camp of Hereford in Texas where he was send during the Second World war (1943) and where he began to paint on kitchensacks (Sacchi). Catrami, Muffe, Neri, Gobbi testimony the primacy of material,

[6] Il consiglio comunale [...] delibera di fare proprio l'invito giá espresso dal Sindaco al Maestro Alberto Burri, del quale apprezza il valore ed il significato del suo messaggio artistico ritenendo che la sua opera per l'ispirazione altamente poetica ed umana più di ogni altra si addice a tradurre, per le generazioni presenti e quelle future le tragedie, le lotte, le speranze e la fede nella terra e nella vita della popolazione di Ghibellina; esprima la sua convinta adesione all'idea progettuale esposta del Maestro Burri in occasione della sua recente vista ed esterna unanimi voti di ringraziamento per la sua presenza e testimonianza delle vicende della rinascita nella nostra città con l'opera che generosamente ha voluto donare [...] Atti amministrativi del Comune di Ghibellina, Delibera del Consiglio Comunale del 25.09.1979.

the dialectic between finite and infinite, form and non-form. Burris natural and emphatic way to identify with material remains a *topos* of his work of art and leads him to Arte Povera and Conceptual Art. His research for organic material, rough linen and the recycling of industrial material reminds one of Joseph Beuys. In Gibellina he decides to create a Cretto in monumental scale (beginning 1987). Similar to a labyrinth or the blow-up of the skin with its rifts constituted of 102 blocs of white cement (height 1,6m) which recover the former village contoured by 12 hectares of green space. The clefts and fissures between the blocs correspond to the former streets. If an ancient inhabitant of Gibellina walks in the inside of the Cretto he is able to recognize the topic position of his house, but he is also forced by the *Verfremdung* of the topical elements to distance himself from the past, to infer new information. His body is set in a new different relation to space. Memory not amnesia. An earthquake memorial which has to last modified by the flow of time passing by. Inspired by temporal and topic dimension of the archaeological sites of Selinunt and Segesta the artist wanted to create a dialectic between the majestic architectures of the past civilisation and the foundation of a imposing site for the hurt population, a dignity elevation for the rural inhabitants. The occupation of a big space like a poured fluid on a smooth surface. Deleuze tells about the nomadic space, 'l'espace lisse', in which the atoms distribute themselves in a sort of turbine-movement. The earth, air and the sea represent this kind of open, smooth space.

> Teoricamente nessun cambiamento, nesuna aggiunta o protesi viene inserita nella morfologia del luogo. In pratica l'artista non ha seguito nessun intervento. Come Picasso, Burri non cerca, trova. Ed è straordinario che questa elementare coincidenza tra opera e luogo possa produrre tanto significato: le macerie di Gibellina, nel momento in cui fanno da supporto all'epidermide del cretto che le ingloba e protegge, diventano virtuale materiale archeologico e consegnano alla posterità il topos chiamato Gibellina; per non dire delle fenditure aperte tra i giganteschi zoccoli di cemento bianco che diventano vere e proprie strade percorribili e ripetono l'antico assetto urbanistico e viario. Il Cretto di Burri allude dunque ad una città nuova, rinata da se stessa dalla catastrofe e in forma di catastrofe, città metaforica ma reale nella sua abbagliante luminosità mediterranea.[7]

[7] Serafini (1998).

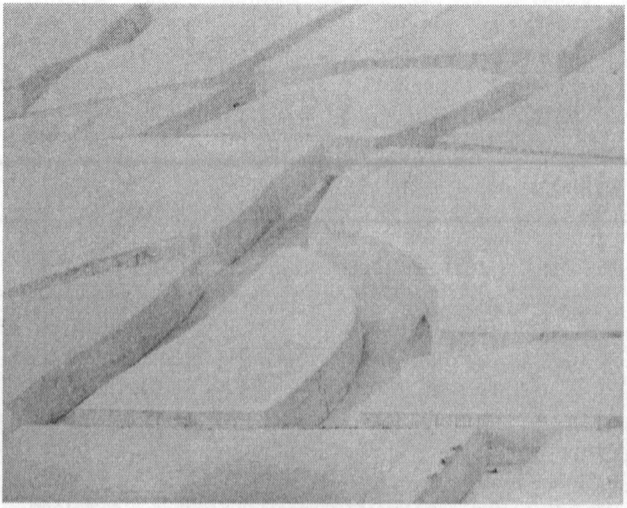

Figure 19

Giuliano Serafini's definition of Gibellina as a remarkable elementary coincidence between artwork and site, coincidence which produces such an enormous semantic density and its definition as a nexus for future generations lead to Michel Foucault's notion of *éterotopie*. One of the constitutive features is the *heterocronie*: the *éterotopie* fully functions in the exact moment of breaking with the traditional temporal belonging. Further it implicates on the one side a system of opening and on the other side of closure: heterotopias are isolated and penetrable at the same time. Another trait which approaches them to Burri's artwork is that they have to accomplish certain functions: in some cases the eterotopic site creates an extraordinary precise, neat and real space in order to compensate with the lacks or imperfections of normal spaces. Heterotopias posses also the power of juxtaposition: one space for different incompatible sites. Diverse traits all applicable to the Cretto.

Conclusion

The architect Giancarlo De Carlo notes:

> A place is an inhabited space. Without space there can be no place, but spaces itself is not enough to make a place because space becomes place if

> and when it its exercised, used, consumed and perennially transformed by human presence. A place is all the more a place the more it is plural [...][8]

In this perspective it could be added that also the gaze constitutes a site. Burri's Cretto, a special example of landscape, was often setting of dramas and plays to which the whole population contributed to. Exploring the urban textum semiotically means reading the visible and invisible lines of cultural stratification. The artistic space is related to or in oppostition to the stratification and represents a reflexive momentum for the definition of the boundaries between urban and environmental space. Burri's Cretto and Gibellina's architectonical and artistic prove how Space Art and Environment can transform or at least play an active part in the interaction between the past of human beings, natural or historical events and the projection into future. *Kunstwollen* grows to *Kunstsein* and finally to *Sein*. Describing, thinking, defining dreams for the appropriation of new spaces.

References and bibliography

De Carlo, G. (2004) *Can non-places be (re)transformed into Places?* In *Domus* **872**.

Eco, U. (1979) *Lector in fabula* Milano: Bompiani.

—. (1962) *Opera aperta* Milano: Bompiani.

Foucault, M. (1984) 'Eterotopie' In *Mimesis* **2**.

Greenberg, R.; Ferugson, B. and Nairne, S. (1999) *Thinking about Exhibitions* New York: Routledge.

Harrison, C. and Wood, P. (1998) *Kunsttheorie im 20. Jahrhundert* Ostfildern-Ruit: Verlag Gerd Hatje.

Holt, N. (1979) *The Collected Writings*. New York: New York University Press.

Mayer, G. and Thoman-Oberhofer, E. (2000) *Landart usw* Innsbruck: Kunstraum Innsbruck.

Serafini, G. (1998) 'Il nome e la cosa, intervento alla mostra-convegno "L'opera di Alberto Burri nel panorama della land-art internazionale"' Ghibellina 10: 8-10

Smithson, R. (2006) 'Essays' On: http://www.robertsmithson.com Viewed 15th May 2006

Volli, U. (2005) Laboratorio di Semiotica. Bari: Laterza.

—. (2003) *Manuale di semiotica* Bari: Laterza.

---

[8] De Carlo (2004: 18).

CHAPTER SIX

NORTHERN PLAINS TRANSFORMED THROUGH NATIVE BOULDER STRUCTURES: ART AND HETEROTOPIAS

THOMAS HEYD

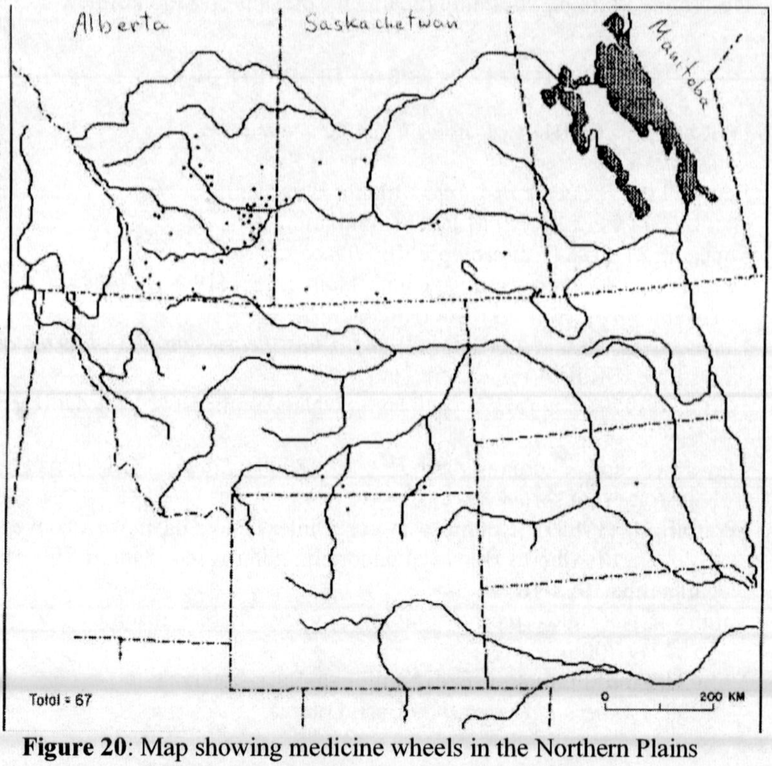

Figure 20: Map showing medicine wheels in the Northern Plains of North America. *Source*: Brumley (1988)

Introduction

'Medicine wheel' is the name given since the late 1800's to a kind of boulder structure found in the Northern Plains of North America. The name has been applied in a generic manner to boulder structures that are relatively similar to the Bighorn Wheel on Medicine Mountain, Wyoming, in distinction to smaller and less complex stone circle structures (such as those often called tipi rings).

Significantly, the name 'wheel', although used in contemporary classification, is misleading if considered as an indication of the image that its original builders may have had. Wheels were probably unknown in the Americas before the arrival of the Europeans. Medicine wheels, moreover, usually are not completely circular; instead of radial symmetry some of these structures have a tendency towards bilateral symmetry (a feature common to all the vertebrates that were known to the Plains Indians).[1]

After a careful survey of the diverse structures usually called 'medicine wheels', John H. Brumley has offered the following definition:

> All medicine wheels consist of a combination of at least two of the following three primary components: (a) a prominent centrally located stone cairn of varying size; (b) one or more concentric stone rings of generally circular shape; and/or (c) two or more stone lines radiating outward from a central origin point, central cairn or the margins of a stone ring.[2, 3]

---

[1] Wilson (1981: 346-7).

[2] But see Nikiforuk (1992: 54) who quotes (no page number given) Michael Wilson saying that "Brumley's definition leaves out a whole variety of closely related stone spokes, circles and simple cairns".

[3] **Note by the author:** For further development of the topic of respect for, and appreciation of, landscapes and nature in relation to native markers in the land, see *Encountering Nature: Toward an Environmental Culture* (Aldershot: Ashgate, forthcoming 2007).

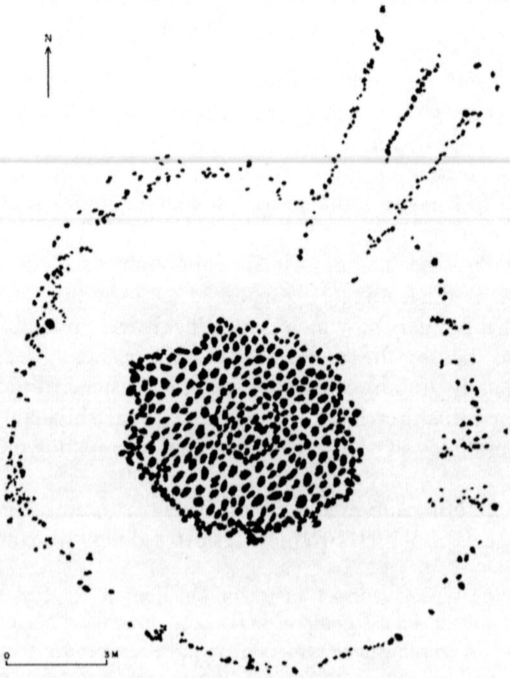

Figure 21: Drawing of Medicine Wheel.
Source: Brumley (1988)

Medicine wheels are often situated on knolls overlooking the prairie, and are mostly found in Alberta and Saskatchewan, and less frequently in Montana and Northern Wyoming. Likely the majority are accretional, having achieved their present shapes through additions over a long period of time. Although some medicine wheels are of very recent origin, others are of great age: the original installation of the Majorville Medicine Wheel, for example, seems to date back five thousand years.

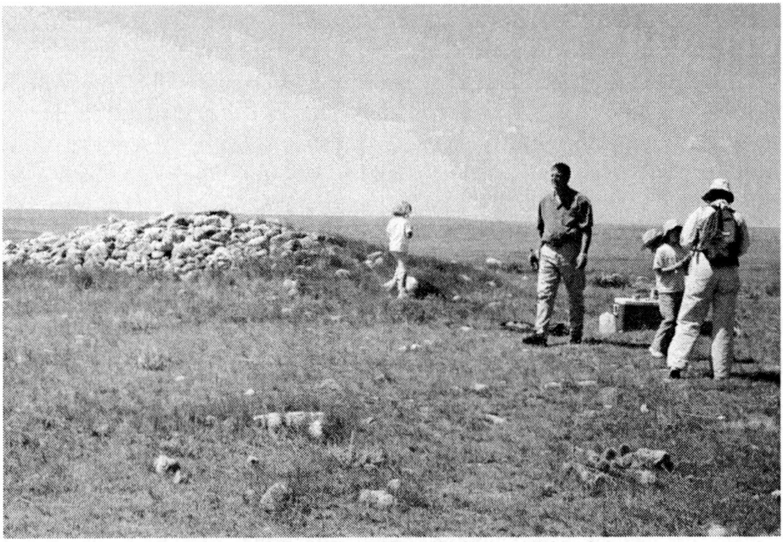

Figure 22: Majorville Medicine Wheel.
Source: Photograph by James Young (2004)

There is great diversity of form among the sixty-seven structures accounted for by Brumley. Although generally not discussed in these terms, medicine wheels often have considerable aesthetic interest. One of the wheels has the shape of a turtle, some resemble nerve cells, while others look like sea urchins. Much as in non-figurative painting, values, such as rhythm in the spoke patterns and spatial relations among the various elements, contribute to their aesthetic effect.

Considerable effort has been exerted in the determination of the meaning or function of these structures. In many cases it has been possible to establish their meaning for the First Nations people of the area. The medicine wheel of Steel (his real name was Ski-matsis, or, in English, Fire Steel), a great Blood warrior, was built in 1940 as a memorial.[4] Moreover, reports from Native consultants indicate that the Big Horn medicine wheel may have been used for vision quests.[5] Nonetheless, the great age of some of these structures, and the fact that there is considerable uncertainty about the identity of the inhabitants of this area at various points in historic and

4 Dempsey (1956: 177-182).
5 Wilson (1981: 337-8).

prehistoric times,[6] leaves the determination of the original meaning of these structures quite unsettled.

The hypotheses posited range from the supposition that the wheels served as 'stone age calendars' to the view that they had a role in the periodic thirst dance (also called sun dance).[7] It is, of course, quite possible that, given the variety of peoples with quite diverse cultures who may have participated in the installation and use of these structures, each wheel had multiple functions.[8]

The combined picture created by the testimony of contemporary Native people about the meaning of these structures for them, and by the archaeological research about their original functions, certainly contributes to the appreciation of these installations as meaningfully inscribed, culturally structured entities; it saves us from the ethnocentric supposition that the Northern Plains may have been cultural wastelands, patiently awaiting the arrival of their branding by feudal ranching and industrial farming practices.

Nonetheless, neither Native nor archaeological accounts fully plumb the significance of these sites for contemporary non-Native peoples in the region. Unreflected, the more fanciful accounts proposed are, moreover, absorbed much too easily by the dominant culture as entertaining but emasculated folklore. I propose that, from the point of view of contemporary mainstream North American culture, these structures deserve consideration as art.

Medicine wheels: Boulder arts

Prehistoric pictographs and petroglyphs generally are described as "rock art", despite the general reluctance of archaeologists to engage themselves in their artistic interpretation.[9] For example, while urging a certain degree

---

[6] Forbis (1963: 9-16).

[7] It is to be noted that deconstructionist theory has thrown into question the very idea that the intentionality of the builders and users of prehistoric sites is something stable that can be found or reconstructed; see, e.g. Davis (1992). Also, see Wilson (1981: 336), who notes that "we cannot dig up ideas". We may assume, however, that we are on relatively safe ground as long as we only seek to establish the *functions* of items, and *not* the intentions of the individuals who made those items.

[8] Wilson (1981: 336). Also, see Bednarik (1991-92: 14) who argues that "the older rock arts were … integrated into [newer] belief systems, [and] reinterpreted...."

[9] But see, e.g. Tilley (1991) who is quite willing to interpret archaeological givens. Regarding the willingness to speak of "Ancient Amerindian Art," and to call

of caution in the interpretation of the petroglyphs and pictographs found at Writing-On-Stone in Southern Alberta, Martin P.R. Magne and Michael Klassen freely and repeatedly make reference to those manifestations as 'rock art'.[10] Similarly, in a paper on southern African pictographs, in which he recommends that linguistic data be considered alongside ethnographic data, J.F. Thackeray frequently refers to the manifestations in question as 'rock art'.[11] Thackeray even notes that "The art has been referred to as 'The San artistic achievement'" without considering it necessary to raise the question whether these manifestations should be referred to as art.

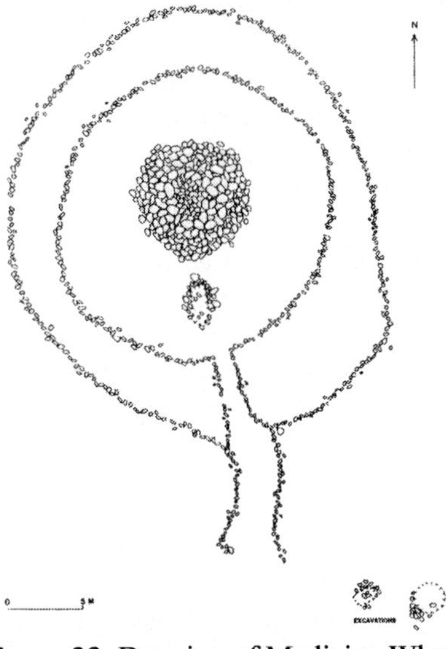

Figure 23: Drawing of Medicine Wheel.
Source: Brumley (1988)

It seems that the appropriateness of calling prehistoric manifestations 'art' becomes subject to examination only when manifestations other than pictographs and petroglyphs are under consideration. Paul S.C. Taçon, for example, raises this question with respect to the legitimacy of calling

"artists" those who fashioned paintings, sculptures, etc. in the Early Americas, see e.g. Kubler (1991). Also see Heyd and Clegg (2005) about the issue of aesthetics and rock art.

[10] Magne and Klassen (1991: 389-418).

[11] Thackeray (1990: 139-44).

certain prehistoric stone tools 'art'.[12] He notes that the reluctance to consider stone tools as art partially arises from the lack of "a cross-cultural, non-biased definition of art." Without claiming to develop criteria that justify the appellation 'art' for any one manifestation, he suggests, nonetheless, that "many forms of stone tools produced over the past 6000 years in western Arnhem Land" be considered as art because these tools "have both aesthetic and symbolic value which influenced their manufacture".

The literature dealing with the definition of art is large.[13] *Prima facie* the range of options is great: some only admit as artworks items created with the intention of making art; others require an art-making tradition as a prerequisite for an object to be a candidate artwork. Presumably, if one settled on a particular definition one could reach a decision on the issue whether medicine wheels should be considered art. Given the ongoing debate regarding the nature of art, having to settle that question would take us too far afield from our purpose.[14] Here I intend to postpone entering that debate, and to limit myself to the observation that, given their aesthetic qualities and their potential for symbolic interpretation (as described above), there is as much reason to consider medicine wheels in terms of a sort of 'boulder art' as there is for calling pictographs and petroglyphs 'rock art'. Medicine wheels, moreover, share crucial family resemblances with certain contemporary artworks in the category 'earth art' or 'land art', which lends further reason to the proposal that medicine wheels be considered art.

Michael Heizer, Robert Smithson, Walter de Maria, and Richard Long have carried out artworks that in various respects resemble medicine wheels. Heizer's Five Conic Displacements, Double Negative, and Complex One, Walter de Maria's Lightning Field, as well as Smithson's Spiral Jetty and Amarillo Ramp, are all located in remote, relatively inaccessible locations. Similarly to medicine wheels, these works introduce marks into the landscape which turn the surrounding land and sky into an all-encompassing stage and mute audience. Some earthworks resemble medicine wheels more straightforwardly than others, though. As has been noted in the art historical literature, some earthworks, such as Heizer's and Smithson's are best seen as extensions of minimalism and conceptual art.

For example, when asked about the role of landscape in his decision to move out of his New York studio into the Nevada desert, Heizer replied

---

[12] Taçon (1991: 192-207).

[13] see Davies (1991 and 2006); Sparshott (1982).

[14] but see Heyd and Clegg (2005).

that he has "no interest in landscape in terms of art ... American landscape art is one thing but my work doesn't have anything to do with that, it has to do with materials. I went to those places for material."[15] Consistent with these statements, the monumental earthwork resembling Central American pyramids called Complex One has been fitted with a 'plaza', located in an excavated depression, intended to obscure from view the surrounding mountain ranges.

Richard Long's land art, in contrast, seeks the integration with the place that it is situated in. Besides considering his programmed walks as art (sometimes he walks straight lines over hills and creeks; sometimes he walks all the roads within a certain perimeter), he also builds stone circles and stone lines in remote locations. He says of these structures that "These works are made of the place, they are a re-arrangement of it and in time will be re-absorbed by it".[16] Similarly to Heizer's works, his stone circles and stone lines depend on local materials; unlike Heizer's works, however, Long's land art also gains its standing from the character of the pre-existing place. Long does not attempt to separate his pieces from the rest of the place as Heizer does by excavating and building obstacles to a free view; Long builds his pieces with the help of, and into, the place. In Long's works one indeed finds strong resemblances with medicine wheels, which also are made from the materials of the place, and which seem to be carefully located in order to suit the preexisting place.

My argument so far has been that it is reasonable to consider medicine wheels as art. Interpretations of tribal artifacts by non-Native people as art have been subjected to severe critique in the recent past, however.[17] For example, with regard to the Museum of Modern Art exhibition *Primitivism in Twentieth Century Art* it has been argued that the categorization of tribal artifacts as art is equivalent to a co-optation by non-European cultures. The co-optation occurs through a process of decontextualization that "transforms cultural objects into contentless forms that can be recontextualized by another culture".[18] James Clifford points out that "Since 1900 non-Western objects have generally been classified either as primitive art or as ethnographic specimens."[19] Either form of "cultural salvage" seems to fragment the integrity of the tribal artifacts in question.

---

[15] Heizer and Brown (1984: 11).
[16] Long (1986: 236).
[17] See Clifford (1988).
[18] Traugott (1992: 42).
[19] Clifford (1998: 198).

The problem at hand seems to be related to the problem concerning 'Understanding a Primitive Society' debated between Peter Winch and Alasdair MacIntyre.[20] The issue in that debate concerns the manner by which the anthropologist is to proceed if, once she has understood something in one culture's terms, she hopes to make it understood in her original culture's terms. If, as is likely, there are concepts in one culture that the other one fails to have, then some kind of salvage operation seems inevitable. This is why it may have been tempting to understand tribal artifacts 'either as primitive art or as ethnographic specimens'. The search for an appropriate, non-ethnocentric approach to tribal artifacts, however, is a challenge that cannot be avoided once the existence of such artifacts has been recognized. Clifford suggests that a non-ethnocentric approach may consist in giving equal importance to the meaning that such artifacts have in their original cultural milieu as they do in non-Native (European, Canadian, or U.S. American) cultures.

When applied to our encounter with medicine wheels, Clifford's suggestion resolves into the importance of seeking and recognizing both the contemporary and the original Native meanings of these structures, while pursuing an understanding of their significance to contemporary non-Native peoples. For a first approximation to the contemporary non-Native significance of these structures, their exemplary integration in the landscape (which contrasts so notably with most of the interventions of our own industrial, exploitation-oriented societies) may be taken into account. Moreover, full recognition of their integration in the land may be facilitated by our conceptualization of these structures as art. On the one hand, recognition as art should counteract the obfuscating tendency to view these structures merely as quaint 'mysteries' and, on the other hand, recognition as art should under-mine the temptation to view these structures in a sterile manner as objects of merely 'empirical' investigation.

In addition to viewing medicine wheels as artworks one may focus on them as sites, as places that transform the surrounding space. In a little studied text Michel Foucault has very productively extended his analysis of practices to the study of our apportionment of space into places.[21]

---

[20] MacIntyre (1970); Winch (1970).
[21] Foucault (1986: 22-27).

Medicine wheels: Heterotopias

Chorographia, a type of descriptive geography first practiced by the Greek Strabo (63 B.C.? – A.D. 24?) and by the Roman of Spanish origins, Pomponius Mela (first century A.D.), may be an approach to places worth returning to.[22] At this point in time most of the Earth's surface has been charted and measured; but, as Michel Foucault points out, places are more than locations in space. That is, places are sites in space that phenomenologically represent "a set of relations ... which are irreducible to one another and absolutely not superimposable on one another".[23] The notion of place is largely getting lost in a society that actively supports the proliferation of establishments such as identical-looking fast food outlets, gas bars, and super-malls, and that largely disorients by images from anywhere, which in turn are reproduced anywhere else, on ubiquitous television screens. Foucault supposes, however, that

> ... perhaps our life is still governed by a certain number of oppositions that remain inviolable, that our institutions and practices have not yet dared to break down. These are oppositions that we regard as simple givens: for example between private space and public space, between family space and social space, between cultural space and useful space, between the space of leisure and that of work.[24]

Interestingly, Foucault speaks of a type of place that may help clarify the contemporary significance of medicine wheels. He says that there are some places that are similar to those that traditionally have been called 'utopias'; these other places he calls 'heterotopias' since heterotopias are real places but "present society in a perfected form, or else society turned upside down". Consequently, heterotopias are "something like counter-sites, a kind of effectively enacted utopia".[25] Heterotopias "have the curious property of being in relation with all other sites, but in such a way as to suspect, neutralize, or invert the set of relations that they happen to designate, mirror, or reflect".[26] To illustrate, the mirror functions as a heterotopia in this respect: it makes this place that I occupy at the moment when I look at myself in the glass at once absolutely unreal, since in order

---

[22] Mela (1988) A related undertaking is the analysis of landscapes in terms of their symbolic meaning or iconography; see e.g. Cosgrove and Daniels (1989).

[23] Foucault (1986:23).

[24] Foucault (1986:23)

[25] Foucault (1986: 24).

[26] Foucault (1986: 24).

to be perceived it has to pass through this virtual point which is over there.[27]

Exemplary are what he calls 'crisis heterotopias'. Crisis heterotopias, as found in traditional societies are "privileged or sacred or forbidden places, reserved for individuals who are, in relation to society and to the human environment in which they live, in a state of crisis: adolescents, menstruating women, pregnant women, the elderly, etc".[28] Crisis heterotopias are places that make reference to the normal condition, and at once call that normalcy into question precisely by giving a place to the extra-normal or extra-ordinary.

Without going into any further detail here, it is notable for our purposes that in Foucault's view heterotopias:

> Have a function in relation to all the space that remains...Either their role is to create a space of illusion that exposes every real space, all the sites inside of which human life is partitioned, as still more illusory. Or else, on the contrary, their role is to create a space that is other, another real space, as perfect, as meticulous, as well arranged as ours is messy, ill constructed, and jumbled[29]

According to Foucault, in the European-influenced, contemporary societies the theater, the cinema, the garden, museums, libraries and barracks function as heterotopias. The cinema, for example, creates illusions that make us question our categorizations of real space; museums and gardens create real space that in their 'perfect' order of things cause dislocations in our perception of ordinary space. In other words, heterotopias serve as places of disturbance; their existence unsettles the regular categorization of our living space.

Given this description it would seem that certain artworks are also examples of heterotopias. Cubist paintings, such as Picasso's *Demoiselles d'Avignon* may create illusions that denounce our apprehension of real space in terms of orderly, frontally perceived images as illusory. Alternatively, rigorously planned Renaissance city squares are real spaces that, in their designed perfection, recall and highlight the messy, ill-constructed spaces in which most of us actually live.

---

[27] Foucault (1986: 24).
[28] Foucault (1986:24)
[29] Foucault (1986: 27).

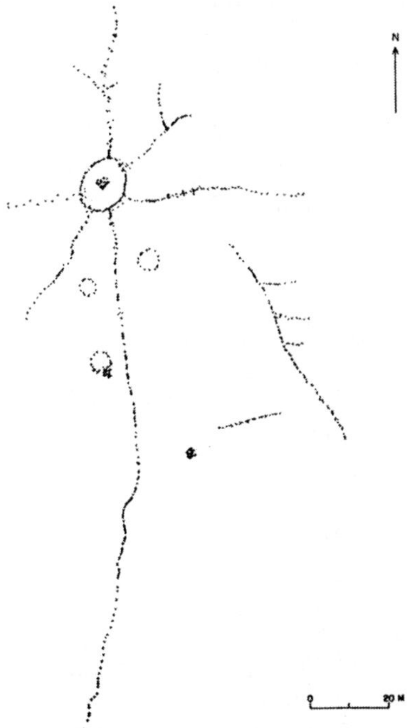

Figure 24: Drawing of Medicine Wheel.
Source: Brumley (1988).

Conclusion

I propose that, insofar as they have the ability to interrogate the interventionist space of our industrial, exploitative contemporary societies, the sites called medicine wheels function as heterotopias. Medicine wheels seem to be, in Foucault's terms, "places outside of all places"; in their resistance to interpretation according to instrumental notions, they return our gaze unto the prairies, turned denatured, overgrazed cattle pastures, and unto the once verdant river valleys, turned into flooded, megaproject water reservoirs (for example, the Oldman River in Alberta). Insofar as medicine wheels are structures built with and into and not against, the land, as are most of our modern societies's interventions in the landscape, they are "a kind of effectively enacted utopia".[30]

[30] A version of this paper has appeared in English as Heyd (2006) and in Spanish as Heyd (2000). A significantly different version of this paper appeared as

References

Bednarik, R.G. (1991-1992) 'Rock art as a Cultural Determinant' In *Survey: Bolletino del Centro Studi e Museo d/Arte Prehistorica di Pinerolo* **7–8**.

Brumley, J.H. (1988) 'Medicine Wheels on the Northern Plains: A Summary and Appraisal' In *Archaeological Survey of Alberta Manuscript Series* **12**.

Clifford, J. (1988) 'Histories of the Tribal and the Modern' In *The Predicament of Culture: Twentieth-Century Ethnography, Literature, and Art* Cambridge: Harvard University Press.

Cosgrove, D. and Daniels, S. (eds.) (1989) *The Iconography of Landscape* Cambridge: Cambridge University Press.

Davies, S. (1991) *Definitions of Art* Ithaca: Cornell University Press.

—. (2006) *Philosophy of Art*, Blackwell.

Davis, W. (1992) 'The Deconstruction of Intentionality in Archaeology' *Antiquity* **66**:334-347.

Dempsey, H.A. (1956) 'Stone "Medicine Wheels" – Memorials to Blackfoot War Chiefs' In *Journal of the Washington Academy of Sciences* 46(6):177–182.

Forbis, R.G. (1963) 'The Direct Historical Approach in the Prairie Provinces of Canada' In *Great Plains Journal* **3**(1):9-16

Foucault, M. (1986) 'Of Other Spaces' In *Diacritics* **16**: 22–27.

Heizer, M. and Brown, J. (1984) 'Interview' In J. Brown (ed.) *Michael Heizer: Sculpture in Reverse* Los Angeles: Museum of Contemporary Art.

Heyd, T. (1998) 'Northern Plains Boulder Structures: Art and Foucauldian Heterotopias' In É. Darier (ed.) *Foucault and the Environment* London: Routledge.

—. (2000) 'Estructuras de Cantos Rodados de las planicies de Norteamérica: 'Arte y Heterotopía'' In *Arkeos* **10**(6):231–44.

—. (2006) 'Contemporary Significance of Northern Plains Medicine Wheels: A Foucauldian Interpretation' In *American Indian Rock Art Journal* 21(3): 205-13.

—. (forthcoming 2007) *Encountering Nature: Toward an Environmental Culture* Aldershot: Ashgate Publishing.

(forthcoming 2007). All drawings are from Brumley (1988), with permission for reproduction granted by the Royal Alberta Museum. The photograph of the Majorville Medicine Wheel is by James Young (12 July 2004), reproduced with permission.

Heyd, T. and Clegg, J. (eds.) (2005) *Aesthetics and Rock Art* Aldershot: Ashgate Publishing.

Kubler, G. (1991) *Esthetic Recognition of Ancient Amerindian Art* New Haven: Yale University Press.

Long, R. (1982) 'Words After the Fact' In R.H. Fuchs (ed.) *Richard Long* New York: Solomon R. Guggenheim Museum.

MacIntyre, A. (1970) 'Is Understanding Religion compatible with Believing? And the Idea of a Social Science' In B.R. Wilson (ed.) *Rationality* Worcester: Basil Blackwell.

Magne, M.P.R., and Michael K. (1991) 'A Multivariate Study of Rock Art Anthropomorphs at Writing-On-Stone, Southern Alberta' In *American Antiquity* 56(3): 389–418.

Mela, P. (1988) *Chorographie* [trans. Silberman, A] Paris: Société d'édition "Les Belles Lettres.

Nikiforuk, A. (1992) 'Sacred Circles' In *Canadian Geographic* **112**(4): 51–60.

Sparshott, F. (1982) *The theory of the Arts* New Jersey: Princeton University Press.

Taçon, P.S.C. (1991) 'The Power of Stone: Symbolic Aspects of Stone Use and Tool Development in Western Arnhem Land, Australia' In *Antiquity* **65**: 192–207.

Thackeray, J.F. (1990) 'On Concepts Expressed in Southern African Rock Art' In *Antiquity* **64**: 139–144.

Tilley, C. (1991) *Material Culture and the Text: the Art of Ambiguity* London: Routledge.

Traugott, J. (1992) 'Native American Artists and the Postmodern Cultural Divide' In *Art Journal* **51**(3): 36–43.

Wilson, M. (1981) 'Sun Dances, Thirst Dances, and Medicine Wheels: A Search for Alternative Hypotheses' In M. Wilson; K.L. Road and K.J. Hardy (eds.) *Megaliths to Medicine Wheels: Boulder Structures in Archaeology, Proceedings of the Eleventh Annual Chacmool Conference* Calgary: University of Calgary Archaeological Association.

Winch, P. (1970) 'Understanding a Primitive Society' In B.R. Wilson (ed.) *Rationality* Worcester: Basil Blackwell.

PART II

ART

Chapter Seven

Sensing Space

Andrew A. Stonyer

Attending the Sensi/able Spaces conference gave me the opportunity to visit Iceland for the first time. Although I was familiar with facts such as, 10% of its area is covered in glaciers, only one quarter of the land mass has continuous plant cover with only 1% of it under cultivation, and there is a significant volcanic eruption every five years, I was not prepared for the spectre of the Icelandic landscape that seems to live and breathe and continually reshape itself. In geological terms, Iceland is a relatively recent country, with a land that is literally the living evidence of the drama of the destructive powers that toil beneath; at its most extreme this might be described as the collision of ice and fire. It is these very powers that shape and punctuate the landscape from the powerful brooding monumentality of its volcanoes and their violent underworld, to the transience of the steam that seeps and cascades from the pores of its land. It is indeed a country of extremes, of white silver skies where the luminosity of the horizon is almost always present during daylight; and below, a land of hard rock, with its pervasive covering of moss and lichen. The former is the very essence of light, where transience, reflection and transparency are the counter play to the opacity and density of the land beneath; and where the polarity of the solstices is defined by the continuous daylight of summer and winter's nocturnal darkness. Yet, the contrasts do not end here; Iceland is the meeting point of the Mid Atlantic Ridge, where the American and Eurasian plates which are moving apart at the rate of two metres a century, leave the drama of a rift valley in between that is a melange of broken rock, bog and water. Almost certainly the drama and recognition of this geological activity is deeply ingrained in the culture and psyche of Icelanders. It is surely no accident that in 930AD, the first Icelandic parliament – the Alþingi, was constituted at Þingvellir, on the edge of the fault line approximately 50 kilometres east of Reykjavík. I would find it difficult to call to mind a site that is more context-specific and more emblematic in terms of defining a country's cultural identity.

Inevitably the significance of the country's geological condition and the visual richness of its atmospheric effects, have an influence on its art. Although the air is possibly the clearest and most unpolluted in Europe, and one can see far into the distant horizon, this horizon is often so mediated by subtle changes in the quality of light; that it is often difficult to ascertain its distance, particularly when subject to the reflections of glaciers. Instead therefore, of the horizon conforming to a keenly defined line where sky meets land, the confluence can be ambiguous and even enigmatic. This ambiguity is further heightened by a landscape that is in the process of formation, it is animate with a skin that in some locations is so warm that it sweats from the geothermal activity beneath. The converse of this of course is the darkness of winter, where sky and land are indiscernible and the horizon is not so much seen as felt, and the perceptions and sensations of each become internalised and synthesised.

For those with little knowledge of Iceland, it would not be unusual to suppose that in the visual arts of Europe's newest island, the sea would be a predominant subject. That this is not the reality may be due to the political history of the country. Whilst there is abundant evidence that Icelanders sailed as far a field as Byzantium and North America over one thousand years ago, such maritime travel and exploration ended when Iceland lost its independence to Denmark and became a subject state with little maritime commerce of its own. Not surprisingly, it is explained, Icelanders looked to the land. As the exhibition 'Confronting Nature, Icelandic Art of the 20th Century'[1] at the Cocoran Gallery of Art in Washington, USA confirmed some of the most significant painting was concerned with the detail of the land, as well as the broader context of the country's landscape. Pre-eminent in this is the work of the painter Jóhannes Kjarval where the actual painted surface makes constant and compelling reference to the texture of the landscape. Paint renders the precarious softness of lichen, in contrast to the unforgiving sharpness and pitted surface of the rock beneath that is subservient only to nature itself, and has an almost palpable reality, indicative of the foreboding powers that lurk beneath. Relief from the powerful gravity of this imagery comes in the transparency of water or sky, which in such paintings as 'Mt Skjaldbreiður 1957 – 62', is given a limited, but crucial part of the composition.

[1] Kvaran and Kristjánsdóttir (2001).

On Vesturey, the north- western part of the island Viðey, that is little more than a five minute ferry ride from Reykajavík, is Áfangar, a sculpture by the American sculptor Richard Serra. As the first port of call for the conference delegates of Sensi/able Spaces in the spring of 2006, it is of interest because of the way it engages our attention, thereby enhancing our perception and sense of space. This is achieved through the sculpture referencing the topology of its location, therefore making it context specific. Áfangar consists of nine pairs of basalt columns, one of each pair three metres high, the other four. These columns are positioned close to the perimeter of this part of the island; the three metre high columns being 10 metres above sea level, and those four metres high, at an elevation of nine metres. Therefore, although the columns are of different lengths; their tops are all the same height. This arrangement of the columns focuses attention on the actual topography of the island, particularly in the immediate vicinity of each of the pairs of columns, as well as the larger landscape traversed from one pair to another. This enhances our awareness of the innumerable subtleties within this landscape, as well as the all pervasive rise to the summit of the hill, that is at the centre of Vesturey. Áfangar; like so many of Serra's sculptures, also has important qualities that are unseen. The unseen is in the geometry of the imagined flat plane that is implied by the matrix of polar coordinates that extend from each of the columns, and cut through the land above. The intersection of the horizontal – the plane, and the vertical – the columns, results in a configuration of a cross, which has obvious symbolic connotations.[2]

One of the essential qualities of Serra's sculpture has always been its physicality and directness, and the skill with which he interprets monumentality. Because he usually works in steel and on a very large scale, his sculpture is indicative of a technological age that relies on modern fabrication techniques. Áfangar however is different, because it is made entirely from basalt it has a monumentality that is timeless in which its 18 columns create a link between the rock beneath and the cosmos above. The first impression of this sculpture is that it was created far back

---

[2] In secular art the configuration of the cross is a frequent symbol that underlies pictorial composition. See for instance the Dutch De Stijl movement which commenced in 1917, where its chief proponent the painter Mondrian, uses the intersection of the horizontal and vertical as the primary element in his pictorial language known as Neo-plasticisme. In his *The Rise and Influence of Rationalism in Europe* Lecky (1913: 190-192) gives a particularly erudite description of the symbolic properties of the cross in respect of the man-vertical, in contrast to the land as horizontal and female.

in time, and may even be timeless having an existence that predates history. One's enduring experience of it is the compulsion one feels to walk from one pair of columns to another, to explore the topology and flora that surrounds each, and to see how the contours of the land change between one pair of columns and the next. On average it takes about two hours to visit all nine pairs of columns.

Sculpture is one of the most difficult art forms, at its most successful it engages the viewer so that one irresistibly walks around the sculpture and in doing so realises that its form, although physically static, is subject to subtle and sometimes abrupt perceptual change. At its most profound this may have an effect where neither the sculpture nor the viewer appear the same to each other again. Áfangar is no exception to this possibility, with its 18 vertical elements. The vertical is the primary element in sculpture and it is through it that we confirm our existence in contrast to the horizontality of nature. Consequently the vertical is irrevocably bound up with the notion of a 'sense of place'. It is therefore no surprise that the Greeks called the vertical—the Gnomen. Gnomen is 'to know' meaning that the object is imbued with knowledge to the extent that it can become a physical, spiritual and metaphorical defining rod. As a consequence, it enabled time to be read by the sun's shadow, and through it people could gain a sense of their own physicality and mortality. Áfangar achieves all of these qualities; through the way it engages our senses it influences our perception of ourselves, our perception of others, and our perception of nature and the cosmos and our relationship to it. This gives us the possibility of experiencing an enhanced 'sensing of space' within landscape, that may only be possible in a country such as Iceland where the landscape is a living testament to the ever present forces of nature.

References

Kvaran, Ó. and Kristjánsdóttir, K. (2001) *Confronting Nature, Icelandic Art of the 20th Century* Reykjavík: National Gallery of Iceland.

Lecky, W. (1913) *The Rise and Influence of Rationalism in Europe* Harlow: Longmans.

CHAPTER EIGHT

LAND AND COLOURS THE ICELANDIC LANDSCAPE AND ITS PERCEIVED COLOURS

MANUELA SILVIA CAMPANINI
AND HRAFNHILDUR SCHRAM

Introduction

> Nature, the countenance of the land, has played a significant role in Icelandic twentieth-century art. At the same time, nature has been an important cornerstone of the Icelander's cultural identity and is therefore not only a visibile presence, but a part of the national consciousness.[1]

In the preface of Confronting Nature: Icelandic Art of the 20th Century[2], Kvaran drew attention to the vital role art plays in the Icelandic national consciousness as well as the profound relationship between nature and national identity in Iceland. During the last decade (1995-2005) a new Icelandic self image emerged; one apparently influenced heavily by the imagery of urbanization and the pressure of scalar models derived from industrialized economies. Iceland seems to have acquired a new identity often related to energy production. President of Iceland (1980-96), Vigdís Finnbogadóttir, during the Nordic Conference on Outdoor Life and Health[3], suggested that considerations are required regarding the process of change affecting Iceland and the consequences that may result from it. Changing the use of a territory involves the re-shaping of the original landscape. Due to the re-shaping process of the landscape its colours are

---

[1] Kvaran (2001: 9).
[2] Kvaran (2001).
A book published for the event which took place in Washington D.C. at the Concoran Gallery of Art (13 October – 26 November 2001).
[3] Finnbogadóttir (2005).

addressed in this paper. The process of re-shaping the landscape impacts Icelanders' sense of space and the way landscape is culturally represented.

This study aims to set out a frame-work for further inquiries extended to the Panarctic region. For this reason it is conceived as an experimental grid/model suitable for critical evaluations.

The research

Studies on colour significance grew during the 20th century, as well the studies on light and electromagnetic radiations. Starting from the 1950s, Lüscher's[4] psychological tests became part of the common diagnostic routine. Nevertheless, depending on human subjectivity, the concept of colour is extremely abstract and not easily expressed. Moreover industrial processes have influenced colours and the standardization of large, but always finished palettes of colours. Nowadays new technology allows the industrialized and globalized world to imagine that everything may be bought or learned by staying at home. This new attitude estranges us from perceiving natural colours in our surroundings. This occurs because we are more and more subjected to influences of images conceived, produced and transmitted from colour palettes which vary in standards from the Munsell Book of Colour[5] to the NCS Standard (Natural Colour System).[6]

Nowadays it is possible to apply the NCS standard to software to define all sorts of standard colours through spectrophotometric measurements. The NCS standard is based on six colours considered fundamental to the human eye: four chromatic (yellow, red, blue, green) and two achromatic colours (white and black). The original aim of our empirical research, which is the basis of this paper, was to investigate possible difficulties in determining colours univocally. Generally speaking, there is a lack of subjective awareness about this difficulty. As a consequence problems stratify in the unconscious and arise in the conscious through industry driven experiences. It is then sometimes puzzling to see colour in its natural environment as opposed to how we have grown accustomed to seeing it through the media.

---

[4] Lüscher (1976).

[5] The book was compiled at the beginning of the 20th century and is used as the standard by Canada, USA and in Japan with little variances.

[6] The system was established in the fifties and sixties by Anders Härd of the Swedish Colour Foundation. NCS system is the Swedish standard n. 55 01 91 02 adopted in many European countries which makes possible to systemise colours to be codified and represented as they are perceived by the human eye and not as derived from a mix of pigments or light reflection.

The two authors, researching environmental sustainability, focused on the human perception of the Icelandic landscape via colours and art. In the research, perceived colours are evaluated and compared to colours as they are represented. The research *in situ* attempts to verify, firstly if colour and landscape as a team, influence colour perception of local residents. Secondly the research attempts to see if the landscape's changing colours impact human activities.

This paper discusses how the relationship between colours and landscape may impacts human activity.

The study is divided into three parts: Part 1 "colours and places", part 2 "colours and Icelandic landscape painters", and part 3 "colours and tourist perception of colours in Iceland". These parts represent studies of three samples of colour conditions. One is on seeing how colour relates to names of places, waters and mountains. Second, analysing colour conditions in paintings by Icelandic landscape painters, and thirdly we surveyed a large group of foreign tourists.

The method

The methodology of the research is empirical-comparative in which both direct observation and data sampling were used. To rationalize the research, Iceland was divided, for all three research parts, into 4 location cells: NW – NE – SE – SW.

Part 1: In identifying how colour relates to names of places, waters and mountains we listed the colour related names of places, waters and mountains from the Touring Map of Iceland (2006) in the scale 1:500 000 and its place names index. Upon processing this data it was possible to create a colour mosaic which included the frequency of colour used in each of the regions.

Part 2: When analysing colour conditions in paintings by Icelandic landscape painters we selected eight painters (four women and four men) amongst Icelandic modern landscape painters and one of their paintings. These are:

Table 1: Icelandic modern painters and a painting selected for the project.

Authors	Gender	Painting
Þórarinn B Þorláksson	M	Þingvellir, 1900, oil.
Jóhannes S. Kjarval	M	Moss by Vífilsfell, 1940, oil
Kristín Jónsdóttir	F	Atlavík, 1945, oil.
Nína Tryggvadóttir	F	Composition, 1947, oil.
Ragnheiður Jónsdóttir	F	Landscape, 1977, oil.
Guðrún Kristjánsdóttir	F	Esjan – mountainside, 1998, oil.
Georg Guðni Hauksson	M	Snæfellsjökull Glacier, 1986, oil.
Húbert Nói Jóhannesson	M	True North, 1996-97, oil.

Part 3: When surveying foreign tourists we distributed 269 questionnaires amongst Italians travelling to Iceland in groups for 6-8 days during the summer of 2005. Of these 269 we received 210 fully completed questionnaires. Aside from the standard information gathered during such surveys (e.g. age, sex, education and length of stay in Iceland) the tourists were asked:

If they had any preconceived colour references regarding Iceland before their trip to the country?

If they had any vision impairments or deficiencies that would affect their colour perception?

To write on a map of Iceland (divided in four quarters) two perceived colours per quarter.

To declare one colour that for them represented the whole of Iceland.

Upon processing the questionnaire it was possible to create a colours mosaic.

This three pronged methodology allowed us to examine the use of colour in art and nature as well as the diachronic aspect of the landscape. Starting from the place names used by the first settlers in the 9th and 10th century it is possible to study the colours perceived by the last users, the foreign tourists, passing trough the colours used by the Icelandic landscape painters. All the statistical data were reproduced on a coloured table where all the expressed colours are organized by quarter (NW, NE, SW, SE) and gender (F, M) connecting the geographical data with the human ones and depicting a map of colours.

Synthesis

We processed the data through two steps. First through each of the three parts of the study, and then we draw a conclusion from the three joined together.

Part 1 *Colours and places*: Of the 3.146 examined names on the Icelandic Touring map, 75 (2,3%) were related to colours, as follows:

Table 2: The frequency of colours per defined quadrant of Iceland.

Colours	**Total**	**NW cell**	**NE cell**	**SW cell**	**SE cell**
Red	18	8	5	2	2
Black	16	5	6	2	3
White	13	4	1	6	2
Blue	11	3	3	3	2
Green	9	2	3	1	3
Brown	8	2	1	-	5
Total	75	24	19	14	17

The most coloured cell was the NW and the least coloured was the SW. The most represented colour was red and the most uniformly distributed was blue. The original colour tied to mapping stresses the mental aspect and leads us to speculate on a somewhat primordial GIS (Geographical Information System).

Part 2 *Colours and landscape painters*: Due to the fact that the selected paintings showed blue and green as prevalent colours, we reorganized the selection of painters as follows:

Table 3: The prevalent colours in selected paintings.

Colour	Authors	Gender	Painting
Blue	Þórarinn B Þorláksson	M	Þingvellir, 1900, oil.
Green	Jóhannes S. Kjarval	M	Moss by Vífilsfell, 1940, oil
Blue	Kristín Jónsdóttir	F	Atlavík, 1945, oil.
Blue	Nína Tryggvadóttir	F	Composition, 1947, oil.
Green	Ragnheiður Jónsdóttir	F	Landscape, 1977, oil.
Green	Guðrún Kristjánsdóttir	F	Esjan – mountainside, 1998, oil.
Green	Georg Guðni Hauksson	M	Snæfellsjökull Glacier, 1986, oil.
Blue	Húbert Nói Jóhannesson	M	True North, 1996-97, oil.

We would like to add a few comments on the colour relevance in the paintings. Þórarinn B. Þorláksson and Kristín Jónsdóttir have often been associated with the colour blue depicted in their works. These usually entail a wide horizon with the mountains of the central highlands, often crowned by glaciers. Blue is also associated with the work of Nína Tryggvadóttir who became Iceland's first woman abstract painter, around 1947. In her abstract works, she drew inspiration primarily from Iceland's nature, where blue dominates. Húbert Nói Jóhannesson in the last years worked specifically with the Icelandic landscape tradition. In the past years, there has been a renewal of Icelandic landscape painting and traditional methods. His works express his relationship to his surroundings, specifically the urban environment.

Addressing specifically the colour green Jóhannes Kjarval, who during summers literally became a part of nature where he lived under the open sky, gradually acquired the status of a national painter in the minds of his countrymen. He drew the attention of Icelanders to the beauty of the lava fields, which cover 10% of the country's surface and also invented the close up of the country. The green moss, covering the rugged lava fields, is a theme appearing again and again and there he emphasizes the texture of the earth's surface.

A combination of the two colours can be seen in the interests of Guðrún Kristjánsdóttir. She has been drawn towards the landscape in SW Iceland, not the least Mount Esja, the city mountain of Reykjavík, the capital of Iceland. She paints slowly, placing one layer after another of oil paint on the canvas, waiting for each layer to dry. The texture of the paintings thus becomes ambiguous, and the tones of the picture vary according to the quality of light, just as in nature itself. In a similar combination of colours Georg Guðni paints the land from a distance, barren and deserted, bringing together sky and earth. He depicts mountains and glaciers, heaths, sands and moors. In the last years he has been concentrating on the lush green valleys of the coastline.

In order to more fully appreciate the inspirations of these selected painters it is useful to arrange the chronologically. Þórarinn B. Þorláksson, Jóhannes S. Kjarval and Kristín Jónsdóttir belong to the first generation of Icelandic landscape painters and were all born in the countryside. Iceland was a peasant society until well into the 20th century. At that time all transportation in Iceland was on horse. By way of these rustic conditions these painters acquired very close contact with nature on their working trips across the country. Nína Tryggvadóttir and Ragnheiður Jónsdóttir belong to the second generation of female artists. They grew up in Reykjavík but spent most of their grown up life as residents of New York. They both visited Iceland often during the summer and exhibited their works there. The third generation of painters Guðrún Kristjánsdóttir, Georg Guðni Hauksson and Húbert Nói Jóhannesson were all born in Reykjavík where they live and work.

In a country like Iceland where the sea, mountains and sky so often meet, cool colours are predominant, particularly all hues of blue. This blueness is tied not only to the objective reality, but becomes an abstract entity permeating the entire atmosphere. Icelandic artists, both representational and abstract, often choose to express themselves in bright, undiluted blue colours.

Part 3: *Colours and tourist's perception*: Of 269 participants, 210 returned the questionnaire: 60% female and 40% male. Participants were mostly middle aged: 31% were 51-60 years old and 23% were 61-70 years old. Just over a quarter of all participants declared they did not match Iceland with a colour before their visit. Those who were influenced by colour said they had been influenced by TV (25%), photos (25%), journals (7%) and other (3%). Only one person defined himself Daltonian (green blind) and one other colour blind.

Looking more specifically at the survey questions. One question aimed at weighting the four frequent/dominant colours. They emerged in this order: green 51%, grey 15%, brown 10%, blue 7,6%. Comparing the question on colour expectations and the question on the representative colour of Iceland, green is always prevalent. Only 30% confirmed it was the colour they originally had in mind. Green emerges 23 times out of 34 in the female group and 19 times in the male group. Of the participants 27% admitted to a preconceived colour and 5% didn't express any final colour. Even if the highest number of participants matches the green colour, it seems they weren't interested in colours of flowers which normally impacts landscape perception particularly with yellow, blue, white and pink colours.

Blue is surprisingly in fourth place even though blue is the colour of water, sky and sometimes the ice. It is probable that tourist perceptions may be influenced by the sun glasses they use. Coloured lenses alter perception of colour blue but only slightly alter colour perception of green. This seems to be the only apparent explanation for a weak perception of blue, because all groups were travelling by the sea at least during two days of their journey, followed by visits to lakes and glaciers. Three groups travelled in June and two in the second half of July. The latter enjoyed better weather conditions. In this case the perception of the blue could have been stronger due for instance to the flowering of arctic lupine.

Conclusions

Mapping the colours perceived in Iceland by settlers, artists and tourists, gives us the possibility to analyze several aspects of human psychology as well the psychological aspect of activities to be promoted and supported depending on the perceived colours. In the same way colours and art have a great importance in the "hosting architecture" which has not been too far from what the external environment evokes. Emphasizing these factors aside from the need of subjective compensations makes it possible to reach well being through a necessary harmony of colours.

The conclusions of our research partially based on artist and tourist experiences may be extended to other fields. It may be helpful to develop a comprehensive artistic system toward the need of harmonizing colour consciousness in the immigration/integration process. At the beginning of the 21st century a new Icelandic colour consciousness may be both significant and evocative for the Panarctic region. Currently this area is less affected by the standardization of colours. Both its populations and artists preserve the archaic and archetypal secret of their survival. The Panarctic natives know how to look at colours as the Icelandic settlers did. As the Arctic is considered a sentinel for climate change it should also be considered that the landscape of the panarctic region is witnessing great change. This change is impacting the landscape's colours and consequently its people and local artists.

During the last decades Iceland has successfully faced soil erosion with massive drainage and forestation programs and is exploiting this reclaimed land. Both the statistical data shown in the research and some ambiguous texture in the painting and in the tones are witnessing a sliding from blue to green. If now the landscape starts to look more green than blue, how long will it take before the Icelandic national identity will perceive this change? And what will it imply for art?

References

Finnbogadóttir, V. (2005) 'In National Parks, Outdoor Life and Health' Nordic Conference, 5-7 May.

Kvaran, Ó. (2001) *Confronting Nature. Icelandic art of the 20th Century* Reykjavík: Oddi.

Lüscher, M. (1976) *Il test dei colori* Roma: Casa Editrice Astrolabio.

CHAPTER NINE

BILL BRANDT AND BRITISH 'SPACE'

ANTOINE CAPET

People interested in photography will be familiar with Bill Brandt (1904–1983), described in the blurb of Paul Delany's biography as the man who visually defined the English identity,[1] but probably for different reasons. It is difficult to pinpoint Bill Brandt's 'specialism'—people with different interests will have come across him via their own specialised interests. Probably the most common route to his work is through the illustrations in books on twentieth century British History, notably in the 1930s and 1940s. There are arguably four major phases in his career which are related to British 'space'. First, what we could call his social reportage, with the familiar contrast between North and South, in the 1930s. Then came the war years—which gave birth to what are perhaps his best-known works, at least in Britain—concentrating on the confined 'space' of the shelters. This was followed by his photography of 'open space' in the British Isles—possibly as a reaction to the confined 'space' of the Blitz years. Finally, around 1950 or so, Bill Brandt turned to other shores—in the literal sense—but he never abandoned British 'space' in the process, even if he gave it a lesser numerical importance in his work.

There is a sort of hinge, I would argue, between phase two and phase three, when his growing interest in British 'open space' came to supersede his interest in British space-with-people, a reflection on the apparent abandonment of the social preoccupations which had obviously impregnated his work so far. It therefore seems worth examining the different meanderings in Bill Brandt's career in the light of one of the strands of this edited volume, spelt out as 'the perception and conception of spaces in the built as well as natural environment', because the various phases in his career seem to offer a perfect illustration of these notions, each in its turn, as applied to Britain, from the 'built environment' which is a source of social strife and therefore human disharmony to the 'natural

[1] Delany (2004).

environment', from which recognisable human shapes disappear, leaving room for the harmony of Nature.

I first came across Bill Brandt long ago thanks to his reportage of the London tube during the Blitz in 1940.[2] This reportage is of course a superb example of the use of 'space' by a major photographer—in this case a confined space which gathers a microcosm of the London population. The best-known shot is probably that of 'Elephant and Castle Underground. November 11th,1940',[3] with its piling up of human shapes in all possible lying positions, except for an elderly gentleman who faces the camera:

---

[2] Though there is some doubt on the actual shots which can really be attributed to him. See the *Museum of London* site: 'Bill Brandt was commissioned by the Ministry of Information to undertake a photographic survey of conditions in underground shelters in London. He began in early November 1940 but was forced by ill-health to give up after little more than a week. Another photographer, who has not been identified, completed the task and a set of nearly 180 photographs was compiled' (http://www.museumoflondon.org.uk/MOLsite/exhibits/blitz/bombed/sleep.htm). See also the interesting commentary of the *Victoria & Albert Museum*: 'After the London Blitz began, Brandt was commissioned to record bomb shelters by the Ministry of Information. His photographs were sent to Washington as part of the British government's attempt to bring the US into the war on the allied side. Cyril Connolly published Brandt's shelter photographs in *Horizon* in February 1942. In 1966 Connolly wrote that '*Elephant and Castle 3.45 a.m.* eternalises for me the dreamlike monotony of wartime London.' Brandt himself recalled 'the long alley of intermingled bodies, with the hot, smelly air and continual murmur of snores' (http://www.vam.ac.uk/collections/photography/brandt/biography).

[3] Reproduced on the back cover of Delany (2004).

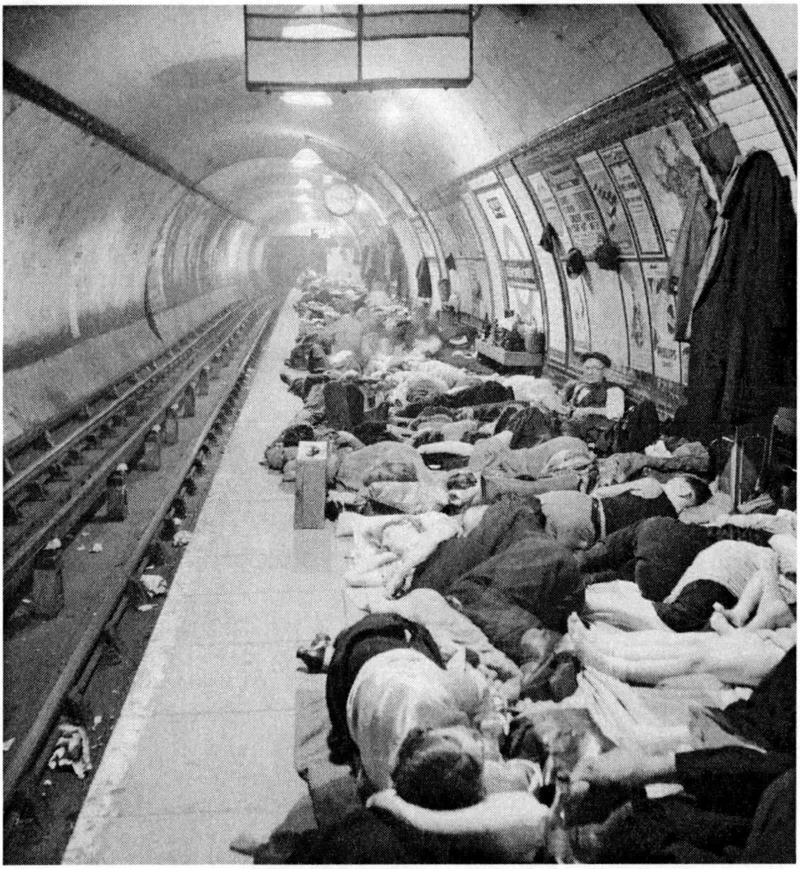

Figure 25: Bill Brandt. Elephant and Castle Underground, November 11th, 1940 (3.45 a.m.).
Source: Photograph courtesy of the Imperial War Museum, London – D 1568.

The same contrast between the huddled sleepers and unidentified onlookers is seen again on a picture taken the next day—or night—at Liverpool Street station:

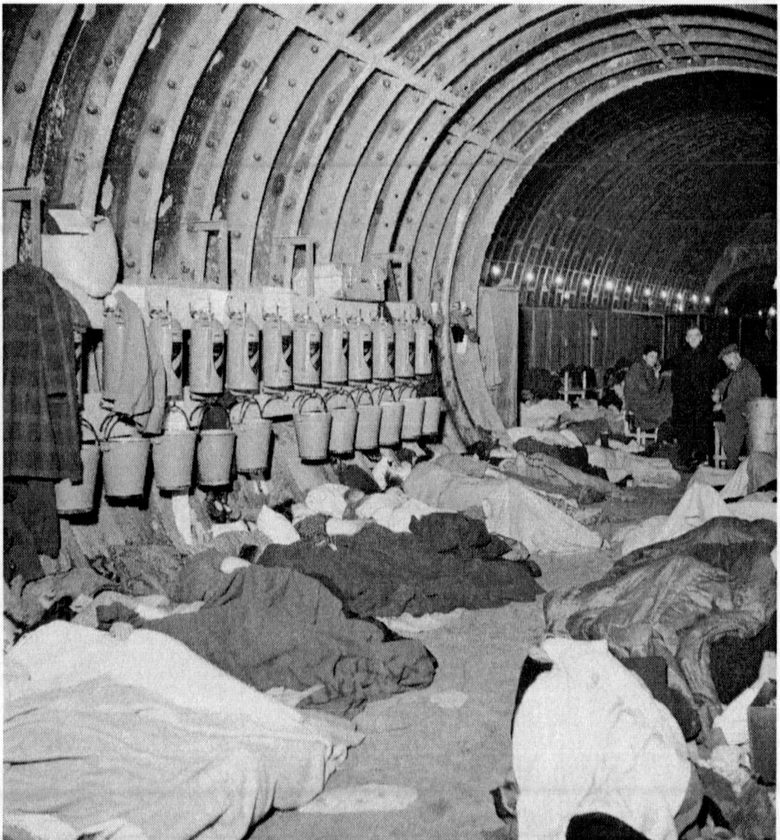

Figure 26: Bill Brandt. Liverpool Street Station, November 12th, 1940. *Source*: Photograph courtesy of the Imperial War Museum, London – D 1580.

Liverpool Street station is of course iconic in itself, as it was the first Tube station to be opened to the public for the night, on 8th September 1940, after the first heavy bombing raids of the preceding night. It is also situated on the border between the City and the East End—hence all the mythology around the social mixing which in fact never took place, as Bill Brandt's composition suggests. The elegant gentlemen in their impeccable coats and the lady with a fine hairdo are obviously middle-class volunteers who give up some of their sleep to bring comfort to people from the East End seeking shelter. The contrast between ordinary Londoners sleeping in the foreground and the protective elites standing in the background is in itself an excellent illustration of how Bill Brandt used space for social objectives before and during the war.

It is arguable that his Blitz reportage was only a continuation of his life-long interest in the poorer classes—the old theme of 'the Two Nations.'[4] In 1936, he published *The English at Home*, in which he contrasted high life and low life. In 'Children, South London',[5] the blurred serpentine background is sharp enough to reinforce the contrast between the infantile liveliness of the characters and the bleakness of their environment. It is clear that the children fill a hostile space, even though they are not yet conscious of it.

In the same collection, we have 'In a Surrey Garden, Cocktails before Dinner',[6] with the same type of composition, in which the human characters fill the lower left-hand quarter of the picture. But here, the pleasure of the guests in elegant evening dress is in harmony with the luscious landscape in the background. Of course, the children did have plenty of 'space' to play—but their 'space', the dingy street, is nothing to compare with the limitless expanse of land, as far as eye can see, which forms the environment of the privileged members of the Surrey party. The life of the rich is defined here as the enjoyment of space which is both beautiful and almost unbounded.

Bill Brandt is fond of these contrasts between rich and poor which play on the contrast between the space allocated to these two classes of society. In 'Drawing Room, Mayfair',[7] which first appeared in his 1938 collection, *A Night in London*,[8] the photographer has deliberately kept the verticals in the panelling which suggest an uncommon ceiling height. The guest in the foreground superbly ignores the spatial privilege which he enjoys and nonchalantly sits on a very low pouf. On the contrary, in 'Window in Osborn Street',[9] from the same collection, the impression given is that of a

---

[4] 'Two Nations, between whom there is no intercourse and no sympathy…The Rich and the Poor'.
Disraeli, Benjamin, Earl of Beaconsfield (1804-1881). *Sybil; or, the Two Nations*. London, 1845.

[5] See: http://arts.guardian.co.uk/gallery/image/0,8543,-10704863641,00.html (viewed: 8th January 2007).

[6] See: http://www.vam.ac.uk/images/image/5020-popup.html (viewed: 8th January 2007).

[7] See: http://www.vam.ac.uk/images/image/5023-popup.html (viewed: 8th January 2007).

[8] Brandt (1938).

[9] See: http://photos1.blogger.com/x/blogger/1669/792/1600/389910/brandt_window_-in_osborn_street.jpg (viewed 8th January 2007).

cruel lack of space, with the top of the basement window almost crushing the heads of the children—the shadow on their hair and foreheads adding to the sense of confinement. The dreary brickwork and the massive scratched masonry lintel which fill the top reduce these human figures to the status of pygmies.

Nature and the burden of life crushing the poor man is of course the theme of one of Bill Brandt's major pre-war photographs, 'Coal-searcher returning Home from Jarrow' (also known as 'Unemployed Miner Returning Home from Jarrow', 1937).[10] The scene did not have to be explained to the British public in 1937. Jarrow was associated with the 'Jarrow Crusade', a march to London organised by the local victims of one of the highest rates of unemployment in the country with the support of their MP.[11] It was also well known that one way of procuring fuel in the mining districts was to scrape the slag-heaps for bits of usable coal. Here the old man bends down under the effort after spending his day filling the bag which hangs on the frame of his bicycle. The slag-heap on the left tells the whole story to the viewer of 1937. Contrary to the Surrey Garden, even if space appears to be boundless, it is evidently hostile—the suggestion is that of barren ground, with meagre grass and heath and a hill which we know is made of industrial waste, not good fertile earth.

Likewise, in 'East Durham Coal-searchers',[12] also of 1937, the bending human shapes on the almost perfect diagonal starting from the top left-hand corner are crushed between the massive blackish slag-heap and the leaden sky—with the incongruous presence of the coal-carrying cables, now idle we guess, radiating from the supporting iron tower. Though composed of only three fairly innocuous elements—the heap, the sky and the tower with its cables—the environment is perceived as hostile for the men. Once more, there is plenty of space for them and for their current activity, but it is obvious that it is not space to be enjoyed.

We could multiply examples, but I believe that the point is made: it is clear that before and during the war, Bill Brandt used space with social preoccupations. How then can one explain that his social conscience

---

[10] See: http://photos1.blogger.com/x/blogger/1669/792/1600/774595/brandt_going_-home.jpg (viewed 8th January 2007).

[11] Ellen Wilkinson, who denounced the scandal of unemployment in her constituency in the famous Left Book Club series. *The Town That Was Murdered: The Life-Story of Jarrow.*

[12] See: http://www.vam.ac.uk/images/image/5099-popup.html (viewed 8th January 2007).

seems to have left him in his post-war work—more and more so as the war years receded?

Admittedly, Bill Brandt had turned to landscapes without a human presence before the end of the war, as exemplified by his famous photographs of Hadrian's Wall in 1943. But his output of the late 1940s definitely constituted an inflection in his favourite themes. The most celebrated work of the period is probably 'Stonehenge under Snow' which made the cover of *Picture Post*, 19 April 1947.[13] Interestingly, the magazine is devoted to the economic crisis which followed one of the severest winters in decades, leading to shortages of all sorts, especially fuel. But then, for Bill Brandt, the appalling weather was a boon. Apparently, he had been biding his time, patiently waiting for Stonehenge to be covered in snow before he photographed it, and the inclement weather which had adverse consequences for the ordinary British population gave him the artistic opportunity which he had long been waiting for. There it is, perhaps, that the hinge which we mentioned earlier is at its most visible. Not only is Bill Brandt's former use of space with social preoccupations totally absent—in fact, if we can use a colloquial expression, 'he could not care less.' Instead of reporting on the war-scarred freezing populace huddled before the few burning fireplaces ('confined space'), as he would probably have done a few years before, Bill Brandt concentrates on the beauty of 'open space' (and without a human presence) for his inspiration.

Other shots increasingly distanced him from the social and economic difficulties of post-war Britain, like 'Gull's Nest, Isle of Skye' (1947),[14] in which the optical effect is a remarkable feature of the construction, with the eggs in the immediate foreground occupying the same absolute space as clouds on the distant background. The use of space has evidently lost all social function, and at a pinch we could say that the hope of a rebirth is transferred to the animal reign. Could that denote disenchantment with mankind after the atrocities of the Second World War? As a native German – though he always tried to pass himself off as British-born, despite his strong accent – it may be that he was vicariously affected by what came to be known among the more progressive Germans after 1945 as 'Vergangenheitsbewusstsein' (consciousness of the past). Whatever the case may be, he seemed to live more and more in a 'never-never' land of

[13] See: http://www.vam.ac.uk/images/image/5013-popup.html (viewed 8th January 2007).

[14] See: http://www.vam.ac.uk/images/image/5038-popup.html (viewed 8th January 2007).

distorted human features, like his 'eyes', for instance Jean Arp (1960)[15] or Jean Dubuffet (1960)[16]—and he seemed to be more and more interested in nudes *per se*, like his 'Policeman's Daughter, Hampstead'[17] of 1945 which has nothing remarkable[18] to make it an immediately perceptible Bill Brandt photograph.

The synthesis between his outstanding use of space and his lasting interest in nudes came with his characteristic beach nudes, typical of what we have called his fourth phase. Here, the link with 'British' space becomes increasingly tenuous: 'Open' space in a given latitude is often deceptively similar and there are pebble beaches with white cliffs on both sides of the Channel. We did not have difficulty recognising his Durham or London backgrounds, but we have to take his word for it when he indicates that some of his nudes were taken on Seaford beach, in Sussex, like 'Nude, East Sussex Coast'(1953)[19] or 'Nude, East Sussex Coast'(1959)[20]. Is there really much to distinguish the background from that of other nudes taken near Nice, on the French Riviera, like 'Nude, Baie des Anges' (1958)[21] or 'Nude, Baie des Anges' (1959)?[22]

---

[15] See: http://www.artnet.com/Galleries/Artwork_Detail.asp?G=&gid=480&which=&-ViewArtistBy=&aid=2968&wid=424325377&source=artist&rta=http://www.artnet.com (viewed: 8th January 2007).

[16] See: http://www.artnet.com/Galleries/Artwork_Detail.asp?G=&gid=480&which=&-ViewArtistBy=&aid=2968&wid=424325375&source=artist&rta=http://www.artnet.com (viewed: 8th January 2007).

[17] See: http://photos1.blogger.com/x/blogger/1669/792/1600/210928/brandt_hampstead_london.jpg (viewed: 8th January 2007).

[18] Paul Delany devotes a chapter to 'The Policeman's Daughter' which reveals that there is a complex sub-text which may be attached to it (viewed: 8th January 2007).

[19] See: http://www.artnet.com/Galleries/Artwork_Detail.asp?G=&gid=480&which=&-ViewArtistBy=&aid=2968&wid=424325384&source=artist&rta=http://www.artnet.com (viewed: 8th January 2007).

[20] See: http://www.artnet.com/Galleries/Artwork_Detail.asp?G=&gid=480&which=&-ViewArtistBy=&aid=2968&wid=424325379&source=artist&rta=http://www.artnet.com (viewed: 8th January 2007).

[21] See: http://www.artnet.com/Galleries/Artwork_Detail.asp?G=&gid=480&which=&-ViewArtistBy=&aid=2968&wid=424325380&source=artist&rta=http://www.artnet.com (viewed: 8th January 2007).

[22] See:

Anybody at all familiar with the sculpture of Henry Moore will be reminded of his reclining figures dispersed in the fields of Much Hadham. If anything, in 'East Sussex Nudes', the 'Britishness' lies in the treatment of the main subject *à la* Moore far more than in the arguably interchangeable 'British' space suggested by the titles. The evolution of Bill Brandt in the 1950s is obviously towards more and more abstraction—the term being taken in the meaning of suggestion. The nude of 1953 is recognisable as a human shape blending into the landscape—not the 'seascape': Delany makes the point that Bill Brandt always carefully shied away from the classic female nude emerging from the waves.[23] The Nice nude of 1958 and the Sussex nude of 1959 present more difficulty; the eye does not immediately identify the shapes in the foreground as human limbs. And the Baie des Anges 1959 shapes are extremely difficult to identify; legs and knees or arms and elbows? Where would the head of the model be—on the right or on the left? With this picture, it seems that we lose our spatial bearings—we literally 'cannot make head or tail' of what we see. What is more, we are at a loss to define the subject of a picture like 'Nude, East Sussex Coast' (1953), because we are unable to tell which is the real 'subject'—the nude or the coast? Nobody can deny that there is a deliberate will on Brandt's part to relate his East Sussex Nudes to 'British' space—but he thereby puts the viewer who knows his earlier use of 'British' space for social reportage ill at ease, because he senses that there is a contrived dimension in all this, as if Bill Brandt wanted to introduce continuity where there is in fact hardly any beyond the reference to 'British' space in the title.

We are therefore left with only one certainty: from the moment he set up in Britain in the early 1930s, Bill Brandt decided to concentrate on British 'space'—as defined in the book, i.e. including 'the built as well as natural environment'. In the first two phases of his British career, it is not difficult for the viewer to follow his intentions, and to perceive his concentration on British 'space', be it the industrial North or Blitzed London. This was predominantly work on 'the built environment'—but built-up areas imply inhabited areas and therefore their inhabitants: hence the 'social' dimension which I have tried to point out. The way that he 'senses' British-space-with-ordinary-Britons is almost invariably presented in a pessimistic light. Then, after the war, for reasons which his

---

http://www.artnet.com/Galleries/Artwork_Detail.asp?G=&gid=480&which=&-ViewArtistBy=&aid=2968&wid=424325399&source=artist&rta=http://www.artnet.com (viewed: 8th January 2007).

[23] 'The conventional glamour shots of women rising from the surf had no interest for Brandt' (Delany, 2004: 241).

biographers have never clearly ascertained, he started to abandon 'the built environment', concentrating on 'the natural environment'—most of it still in the British Isles. The way that he 'senses' British 'open space' or space with only 'abstract' fragments of human shapes on his East Sussex beaches also shows his social pessimism because it can arguably be construed as a form of escapism from the 'real' British space, used by the 'real' British society. While there is continuity in Bill Brandt's uninterrupted exploration of British space over thirty years, and in his fundamental social pessimism, there is an evolution in his description of British space which seems to reflect his gradual aloofness from 'real' people in 'real' society. For Bill Brandt, the sensible/sensable spaces of Britain yield different 'sensations' whether they are full of people (British society is negative) or empty of people (British nature is admirable).

References

Brandt, B. (1938) *A Night in London* London: Country Life.
Delany, P. (2004) *Bill Brandt : A Life* London : Jonathan Cape.

CHAPTER TEN

THE CHANGING SENSE OF A SOCIAL SPACE IN RELATION TO A DEVELOPING EPHEMERAL ARTPIECE

MICHAEL SHIELL

Since the late 1960's the increased significance of context within the creation and exhibition of artworks has ensured that a greater emphasis is being placed upon the physical, social and mental space that works now occupy. Permanent public artworks have been seen to change a society's sense of a given space. Impermanent works that have a definitive installation and de-installation timeline have also been seen to have a similar effect. Ephemeral environmental artworks however do not possess this same definitive beginning and end.

Unlike permanent, solid works designed to withstand the rigours of time, ephemeral environmental works exist within time's cycle. They are not intended to be a massive imposition within a space but rather an element of that space that will pass. This brevity of life when coupled with a lack of any formal de-installation process means that the works departure is more like the gentle passing of a life. Unlike the finality of removing work from a site, devolution allows the work to evolve into something else. In this case the intent of the artist and thus the original work may be lost but the constituent elements of the work can remain. This also allows for the continued evolution of society's connection to that space through that work.

Illustrative of this I will review the developing sense of connection with a specific space that has grown with the devolution of an ephemeral environmental artwork in western New South Wales, Australia. Begun in 2001, the work entitled 'Bone Circle / Bone Fire' has considerably devolved. With this devolution however there has been a development in the significance of the space for the small community that uses the area.

Discussion

In terms of understanding art's potential for the transformative experience of space into place there is an incumbent need to better understand the artwork's relationship to its surrounding. Traditional outdoor works tended to impose themselves upon areas in such a way as to become the focus. Prior to the 1960's artworks inhabiting outdoor spaces generally did not see the specific contextual issues relating to the site as needing to impact upon the work. The art piece was a stand alone autonomous entity. The space it inhabited was merely its surroundings. Works that have been produced more recently however are more likely to take into account at least some of the contextual issues that are applicable to the given space during a works creation.

This earlier attitude of the artwork being the focal element while the outside world tended more to distraction was also reflected the apparent attitudes of the gallery world. In Brian Dougherty's seminal text *Inside the White Cube: The Ideology of the Gallery Space*,[1] he explores how the physicality and functionality of the gallery space framed peoples' understanding of the art experience. He discusses the space of the gallery as possessing a sanctified quality similar to that of a church or a cathedral; a reverent space where silence is observed and the artwork is exhibited and seen in a highly codified and constructed manner. More poignantly however, he also wrote that the physical space of the gallery was created in such a way as to exclude the outside world. Nothing from without is allowed in that may be a distraction to the sanctity of the space and the works on display. With this in mind it is clear why the contextual significance of outdoor spaces did not impact on the works that would inhabit them. These works were not an interaction with space but rather an imposition that forced themselves to be the overpowering focus of that space.

Towards the end of the 1960's the predominance of a single driving style behind artistic development appeared to shatter. At this time we saw the beginnings of Conceptual Art, Arte Povera and Land Art to name but a few of the new directions. Many of these new directions did share varying degrees of overlap; a rejection of the supposed preciousness of the art object, a greater consideration of the role of process in the works creation and a varying degree of rejection of the gallery structure. Arguably the strongest initial rejection of the gallery system was to come through the early American Land Artists. Their distaste of the present system and the

[1] Dougherty (2000).

power it wielded was probably best categorized when Michael Heizer said "The museums and collections are stuffed, the floors are sagging but the real space exists".[2] They purposefully sought locations well outside of the usual artistic sphere as they headed for the Western Deserts of the United States. They further turned their backs on the galleries as they created works that were so massive and tightly bound with their site as to be uncollectible. Beyond their rejection of the gallery system and many of its underlying principles the development of their art forms also began to signal a change in the role of space in the consideration of art. As Michael Heizer stated "the intrusive opaque object refers to itself. It has little exterior reference. It is rigid and blocks space. It is a target. An incorporative work is aerated, part of the material of its place and refers beyond itself".[3] This statement appears as a conceptual precursor to the development of the more recent thinking such as that of site-specificity. Although his sentiments did not overtly speak of a need for a relationship between the work and its site it did flag the beginning of a coming change.

Although these fledgling concerns for site, which would later develop into the issues of space and place, appeared to influence the early Land Artists it is debatable as to whether their monumental results were particularly mindful of their sites. Indeed one journalist remarked that "earth art, with very few exceptions, not only doesn't improve upon its natural environment, it destroys it".[4] The flaws in this sentiment however are twofold. Firstly, beautification was not a goal of these works. Their motivation was driven by a desire to interact with the land not impose the judgement of beauty. Secondly, these massive works were produced at a time when public concern for environmental issues was only just beginning through the growth of the environmental movement. In this regard if some of these works are judged by today's standards they do appear to work against the site rather than with it. In contrast a similar earth-bound practice developed almost simultaneously in Europe but with distinctly different results. The works of artists such as Richard Long and Hamish Fulton were considerably more intimate and personal. In many ways it is the work of these English artists that share a direct conceptual link to today's ephemeral environmental art practices. One aspect of commonality shared between these two differing working methodologies that also influenced other artistic practices was the increased emphasis being placed on the context of works.

---

[2] Beardsley (2006: 13).

[3] *Ibid.*

[4] Auping (1977: 1).

The heightened consideration that is now being given to the context in which a work is being seen has ensured that a greater emphasis is being placed upon the physical, social and mental space that works now occupy. The context has become a primary consideration in the relationship between an artwork and its site. In an attempt to define this relationship Stefanie Ross wrote that artworks could be understood to exist along a continuum. She defined various categories along that continuum from site dominant, through site adjusted, then site specific and finally to site conditioned and determined. The progression through each of these categories meant a stronger bound relationship between the artwork and its site.[5] This understanding of site-specificity does assist in our comprehension of arts potential for the transformative experience of space into place, because works that are more strongly grounded in their site can act as a conduit to the formation of a relationship with place.

One sub-branch of modern artistic practice that owes a lot of its development to the early Earth and Land Art practices is that of Environmental Art. Although not bound by a standard set of codified principles this movement generally is concerned with interaction with the land rather than merely its representation. Through the sited artworks of various artists such as Andy Goldsworthy, Richard Long and Nils-Udo this practice has also defined a different type of relationship to site that is more than just temporary. The affiliation is greater than that of a static object and its place in the land. By re-considering art as a process of interaction rather than solely by the objects it produces, the longevity of works can also be considered as an issue of context.

With regard to environmental artworks the physical connection to the context of a place is self evident. Of particular interest however, is the more subtle connection of ephemeral environmental artworks to the concept of time within the context of a site. The creation of permanent works is an object-driven practice that seeks to create unchanging forms that exist outside the effects and cycle of time. Temporary works such as installation art are still an object-based practice but they recognise the cycle of time. Therefore these works have a definite and limited time for which they exist. The installation and de-installation processes are clearly marked and the boundaries of the work are well defined. In contrast to this there is another creative practice that incorporates the significance of time without the strictly defined parameters of an installation and de-installation phase.

[5] Ross (1993: 175).

Ephemeral environmental artworks may appear similar to other temporary art practices yet they also have some very distinct differences. As a process-driven practice ephemeral environmental artworks are concerned with a greater length of artistic interaction over time. As the name suggests ephemeral artworks not only exist for a short period, but are also a living process. In contrast to temporary artworks that have distinct installation and de-installation phases that are externally applied, ephemeral artworks devolution occurs as a result of internal mechanism. Most commonly these mechanisms are a combination of the material of the work, the site in which it is created and the affects of nature and time. As these mechanisms are internal to the work and not activated at the hand of the artist, the longevity of the work is less defined. Also in contrast to temporary works whose de-installation is harsh and sudden, the devolution of form that occurs with ephemeral works is more akin to the gentle passing of a life.

Ephemeral environmental art by its very nature incorporates a level of ambiguity. It is a process driven creative practice that places no emphasis upon the creation of a permanent form. This reorientation towards the creative process rather than the product means that to comprehend the artwork the viewer must understand that any form that it creates is a function of time. The comprehension of the existence of an artwork however is not easily defined because these works incorporate the devolution of their form as part of the interaction. Therefore even though the form of a work may be breaking down, it still exists as part of the greater art process.

As a result of ephemeral environmental artworks' incorporation of change in the forms it creates, it allows viewers of the work to return and experience both the work and space anew. Unlike a static artwork that potentially may be comprehended in a single viewing, artworks that incorporate change necessitate multiple visits to comprehend a works evolution and devolution over time. Although this series of visits does not guarantee peoples' perception to change from space into place it does allow a greater length of time, and volume of experience in which this change can occur.

The temporal boundaries around these works are not the only margins that can appear blurred. Although these works are an interaction with specific sites, these sites are generally not bordered by any physical boundary. This lack of demarcation defining the works existence is only heightened as the work de-evolves. Its material constituents are still present, however they may have crumbled and fallen, but this destruction was always an intended element within the entire process therefore it still exists within the understanding of what is the art.

The effect of this ambiguity as the work de-evolves can impact the social perception of the space. Unlike temporary artworks whose de-installation is definite, ephemeral artworks devolution means that their presence still lingers even after their form has deteriorated. The lingering remains of an ephemeral piece appear to extend the works capacity to transform peoples' perception of space into place. In contrast to the finality of removing a temporary work from a site, devolution allows the work to evolve into something else. In this case the intent of the artist and thus the original work may be lost but the constituent elements of the work can remain. This also allows for the continued evolution of society's connection to that space through what remains of the work.

Due to the short life span of these types of works they also incorporate a secondary process of documentation. This recording is not done as a means to extend the life of the work, as this would be contrary to the original intent; rather it is done for the purpose of sharing the memory of what has passed. This documentation can then play a continuing role in the ability of the work to transform peoples' perception of space into place.

Case Study

Background

The Science Faculty through the University of Ballarat provides their students with in-the-field training through an annual excursion to a research property in western New South Wales, known as Nanya. Since 2001 these field trips have also incorporated an artist's residency programme that allows for two emerging artists to experience the land of Australia's arid zone. I was fortunate enough to participate in this artist's residency in its inaugural year. The initial appeal of this opportunity lay in its sense of immersion in a unique and altogether unexperienced environment. Additionally the potential for gaining some understanding about how others understand their experience of environment was a considered outcome. In reality the experience was a great deal more than that. The work that I created during this residency was to have an ongoing affect beyond anything that I could possibly imagine prior to the experience.

Before applying for the Nanya residency I had focused upon creating site specific ephemeral artworks in familiar environments. In so doing, I was working with areas that were charged with personal history and significance. I felt that this lack of a personal narrative in response to that land was of some concern; however any anxiety I may have felt turned out to be ill-founded. The breadth of scientific knowledge allowed for an augmentation of my own experiential understanding, which was lacking in this land. Although this scientific basis created a solid foundation it remained merely a point of origin in terms of the creative process. It formed the beginning dialogue between myself and the land, however the final resolution for the work was far removed from the clinical, objective results of scientific observations.

Primarily my work focused on the need for and lack of water within this landscape. As part of the arid zone, this area of Australia was at that time experiencing its fourth year of drought conditions. As a result of this the signs of water deprivation were evident upon this scorched land. The lack of water created a very real and tangible pressure to the survival of both plant and animal life in this area. Survival pressures within specific environments also shared a degree of overlap with an ongoing interest I had in shelters as sites of relative warmth and safety along the length of a journey through the land. Far from being a point of disjunction between an organism and its environment, a site-specific authentic shelter can speak of the environment and the organism's relationship to it.

During the beginning of the residency I was less concerned with creating works as I was with adjusting to the space. Absorption, reflection and the subsequent creative expression is a journey that requires both experience and time. During this period I gained both an objective understanding of what I was seeing through the scientific information being provided, while also being aware of my subjective responses to both the sites and materials of this land. The overall sense of space is one of great openness. The minimal undulations among the vast open surroundings gave a sense of great distance. Within this expanse I sought areas of slight depression, for had there been water it would have flowed and pooled here. These gentle recesses became the intimate sites for the works.

Working around this framework I made a number of smaller artistic interactions with the land. In each of these pieces the materials were sourced very near to the site where the works were then created. However it was the final work done during the residency that is of the greatest interest; not only for the work itself but also for the ongoing transformative result that it has had upon the scientific researchers and students who continue their own work at the property.

Bone Circle / Bone Fire

As the final work that was done during the residency the piece entitled Bone Circle / Bone Fire was the most resolved. At a superficial level the work could be understood as an interaction with the land in general, particularly in a land of such vast openness. In reality however the conceptual basis was drawn more generally but the work was a response to the small site in which it was created. The primary material that was used for the work was sun-bleached animal bones which were collected close to the depression where the work was situated. These bones were indicative of the harshness and aridity of this land. Its prevalence was also due, at least in part to the drought conditions affecting this part of Australia.

In deciding specifically on which slight depression to utilize for the work I decided on one that was close to the main animal drinking area on the property. Being almost completely dry it meant that many animals had perished in this area. I collected all the bones I would require from within a 150 metre radius of the site. I then swept the loose sand out of the depression with the intention of highlighting the cracked clay of this dried sunken hollow.

The bones were then arranged in a circular form within the depression. The purpose of working in this circular format was in direct response to the place of water in this land. As water flows along these vast open areas it gathers the loose debris and plant material that is found on the land's surface. This debris moves with the flowing water to collect in the puddles that settle in the depressions on the land's surface. The debris moves to the waters edge in response to the action of wind and water. As the water evaporates the residual plant material is left in a pattern of circles and lines depending on the nature of the flow. In effect this patterning could be seen as a tiny topographical map as the debris traces the contour of the greatest height that the water reached.

The work remained in this form for the rest of the day but was destined to change after the setting of the sun. As part of the conceptual basis behind my environmental art practice is to create ephemeral works that have a changing life, it is not uncommon for my pieces to have a number of developmental stages prior to its own natural devolution. The next phase was to incorporate a cairn-like structure in the centre of the circle that would be internally lit under the darkness of the night sky. Using the remaining bones I began to build the small cairn skyward.

Beginning from the basic concept of the shelter as a place of respite and nurture in a harsh landscape I decided to make the cairn hollow, which also meant that the work was more fragile and susceptible to environmental conditions. This fragility was a vital element towards the honesty of the work as it reflected the tentative balance of life in this land. Since this work was also inspired by the dynamism of natural processes it would seem fraudulent to then create works that were dense and long-lived.

During the residency the moon in the night sky was almost full. This meant that working by moonlight alone was not difficult. Due to the extent to which the bones had been bleached by the suns rays they appeared to glow under the radiance of the moon. Under these conditions the sun-bleached surface of the work created an eerie glow that seemed to resonate within the landscape. There was another stage however, during the works evolution that was to radiate more light.

During an earlier work I had observed that the bones held with them a sense of foreboding. The presence of death seemed to linger on. The challenge was then to use bone material as the basis for the work, but also to incorporate another element suggestive of regeneration and life as it applied to this environment. Fire is vital to the regeneration and bio-diversity of plant species, which then support animal species in this area. I therefore opted to include this as the additional element within this work.

Working carefully so as not to topple the fragile bone cairn a selection of dry grasses and small sticks were placed inside. The incendiary materials were then lit. Fire breathed a new life into the work. It was no longer a static form. The flame flickered and danced in response to the gentle breeze passing through the openings within the bone cairn. This dynamic movement brought a play of light and shadow to the bone circle. In turn the cast shadows danced around the ground, creating new energy and life.

Eventually the warmth and light of the fire died, leaving only the structure. This structure remained intact for the rest of the residency.

Post Residency

After the completion of many artist residency programmes, the artist walks away in the knowledge that the experience had is completely over. Very rarely would an artist get a second chance to revisit the same site through the residency programme. At the time of leaving Nanya I also believed that to be the reality of that experience also. I knew that I would pine to return to the silent, peaceful spaces of this harsh arid land but the opportunity would not come again as the programme was designed to give different emerging artists the opportunity from year to year. Therefore when I left the intact work Bone Circle / Bone Fire it was with a longing as I knew that I would not have the opportunity to see the manner in which it returned to nature.

In September 2004 (three and a half years after the initial residency) I had just begun postgraduate studies through the Arts Academy at the University of Ballarat. The Nanya residency programme was calling for proposals from the latest group of emerging artists. I did not apply as I had already had my experience and I did not wish to deprive other artists of theirs. Once the artists were chosen I was approached by the organisers who asked if I would like to attend again. As a postgraduate researcher I could do my work at Nanya without depriving others of the opportunity of attending the residency. Although a considerable amount of time had passed I was still very keen to see what remained of that work. Prior to leaving however I was to learn a great deal more about the way in which this work had affected others.

One of the most visibly evident indicators that the work had impacted others and their experience of this place became notable during a pre-trip meeting. After the initial residency the Science Faculty had purchased a very large photographic reproduction of the work, which had since hung in the main office of the school. While attending a meeting at the school I noticed that beside the large photograph were a number of smaller ones that been taken over the passing years. Upon enquiring I was to learn that the students had independently recorded the works devolution. As the students who attended these excursions generally changed every two years I was surprised that there was this level of continuity. During the 10 hour bus ride to the property I was to learn in part how this transformation had occurred.

Within the environmental management course at the University the excursion to Nanya is a requisite part of the programme. Students in their second and third years of study visit the property to learn through doing their own first-hand research. Over the years staff had begun to use the photographic record of the work as a tool to give the incoming students a sense of the Nanya experience. This image was also used within the Science Faculty to be indicative of the role of the artist residents and suggestive of the types of outcomes they may produce. As a result of this exposure students became highly accustomed to the work, however this was not the only exposure they received to the work prior to the excursion.

Other than the students, a minimum of four staff from the school also went on the excursion. It was these staff members that provided the continuity for giving the students a basis for the work prior to their arrival. As I was to learn from the students one particular member of staff spoke about this work with great enthusiasm. In his retelling of the works creation he imbued the story with so much fervour and life that the students' explanation sounded more like mythology than a process that I had actually lived. This form of oral history may lack the objective observation and certainty of detail that a definitive record produced at the time of the work's creation may have had, however its ability to be retold brings with it a sense of life. Details can be blurred a little, potentially elements may even be lost, but if the speaker's enthusiasm means the overall event still lives and appears relevant then maybe the cost is negligible. With regard to ephemeral works that incorporate the element of change as a vital part of their identity then it may well be relevant that oral history be used as a form of documentation. The work itself is born out of living process therefore a living record such as oral history and story telling would also be relevant.

During this return trip I noted a further two things of interest with regard to this piece. Firstly the manner in which the work had devolved was particularly interesting in regard to its initial inspiration. The work was inspired by not only the aridity of the land but also water and the manner in which it flows through a landscape. The devolution of the work had occurred in a direct relationship to this inspiration because the flow of water and the movement of the wind that has changed the work. Over the three and a half years since my previous visit the property had received some rainfall. This rain coupled with the effects of wind had brought loose sand and grass seeds back into the depression where the work stood. As had inspired the work, this flow brought that material to rest at the edge of the work. As the water evaporated the seed began to germinate. What remained then upon my return to the site was a shadow, a living ghost. A reminder of what had once been. The second point of note was the fact that the site of the work had now increased in apparent significance. Prior to the work this site was of no greater significance than any other area on the property. However over the previous years a visitation to the site had become a fixture in the excursion timetable. Upon seeing what remained of the work I quickly realised that it was not the work that drew them back but rather the place. The work became the conduit through which the experience of this space became one of place.

Conclusion

The unique relationship between ephemeral environmental artworks and the places they inhabit is distinctly different from both permanent and temporary art. As a process-driven art form its concern lies in the interaction with, and the formation of a relationship with a place. Since it is no longer the object but rather the process by which a works existence is measured this leads to a level of ambiguity about where the artwork begins and ends; both spatially and temporally. The spatial ambiguity occurs as a result of a lack of delineation regarding the physical boundaries of the work and the environment. Unlike the edges of a photograph these works do not have strongly marked limits. The temporal ambiguity is brought about as a result of the artistic interaction occurring over a length of time that also incorporates the devolution of the form as part of the artwork. If the conceptual underpinnings of the work intend for its devolution to be integral to the understanding of the piece then the form is no longer central to the works comprehension. One result of this uncertainty is that the work becomes an integrated element of that place. This tie between the work and the place can allow viewers to form a stronger bond to, not only the changing work but also the area it inhabits. In this way the work can act as a conduit to the greater experience of the place in which it stands. Although the artwork is not guaranteed to transform peoples' perception of space into place, it can assist with the formation of an ongoing relationship to the area.

References

Auping, M. (1977) 'Michael Heizer: The Ecology and Economics of Earth Art' In *Artweek* June 18th

Beardsley, J. (2006) *Earthworks and Beyond* Abbeville Press Publishers: New York.

Dougherty, B. (2000) *Inside the White Cube: The Ideology of the Gallery Space* University of California Press: San Francisco.

Ross, S. (1993) 'Gardens, earthworks and environmental art' In S. Kemal and I. Gaskell (eds.) *Landscape, Natural Beauty and the Arts* Cambridge University Press: New York.

CHAPTER ELEVEN

MOVING MOUNTAINS: THE VIEWING OF NATURE AND THE NATURE OF VIEWING IN NÓI ALBÍNÓI[1]

SHAUNA LAUREL JONES

Introduction

How do we sense the spaces of the natural environment, and how do we make sense of what we sense? The 2003 film 'Nói Albínói' by writer and director Dagur Kári Pétursson presents a character whose struggle with this very question stems from the contradictions between the way that he psychologically perceives the landscapes of his literal surroundings and his fantasies, and the way he is physically connected—or disconnected—from them both.

In visual art and in the general human imagination, the aesthetic concept of "the environment" and landscape is predicated by the

---

[1] Many thanks to Professors Laurie Monahan and Ann Bermingham of the History of Art and Architecture Department and Professor Edward Branigan of the Film Studies Department at the University of California, Santa Barbara for their invaluable input and contributions to my analysis of *Nói Albínói*. I am also indebted to Dóra Hlín Gísladóttir and Hálfdán Bjarki Hálfdánsson for hosting me in Ísafjörður and Bolungarvík and taking me on an unforgettable airplane tour of the Westfjords; to Arngunnur Ýr, Solla Sigurðardóttir, Gunnar Gunnarsson, and numerous friends in Reykjavík for countless hours of conversation and informal lessons in all things Icelandic; and to Hafþór Yngvasson, Director of the Reykjavík Art Museum, for encouraging me to look at visual representations of both intimate *and* distant nature in Iceland. Portions of this essay were presented in the Music and the Moving Image conference at the University of California, Santa Barbara, 14–15 January 2006, and I give thanks to the conference organizers for inspiring my analysis of the complex relationship between music and landscape in *Nói*. Finally, I would like especially to express my gratitude to Dagur Kári for his generosity in conversing with me about his film.

distancing quality of the urban. This physical and psychological separation of humans from "nature" has given rise to both our sense of control over and exemption from the environment as well as our aestheticization of it. But this lens of distance is a distorting one. Just as artists have played a significant role in shaping our view of the non-human environment, they are now positioned to critique that construction and the grave consequences it has effected; after centuries of looking at landscape in artistic rendition as static and to be seen from a fixed, distant perspective, newer representations—especially in the moving image—might impart the experience of landscape as an embodied and peripatetic one. And beyond the conservational significance of reimagining our aesthetic, psychological, spatial and material relationship with the environment, simultaneously reconsidering our relationship with art objects or films that portray nature takes us a step farther into the question of materiality itself as a relational quality.

Through the narrative vehicle of a teenager's fantasies of escape from his alienation in the Icelandic Westfjords, Pétursson's Nói Albínói challenges the idea of landscape as distant and immobile, both through the eyes of the viewer and the protagonist Nói himself. Pétursson's filmic space is often claustrophobic in its idiosyncratic uses of light, colour, and *mise-en-scene*; this creation of uncomfortable space reinforces Nói's conflict between his imagined experience of remoteness of a tropical landscape that he fancies as a source of psychological relief and his perceptual experience of intimacy with the same palm trees in his ViewMaster toy, on his wallpaper, even on his father's Hawaiian shirts. The film's score heightens this psychological/experiential friction and suggests an active relationship between music and landscape. And just as the actual Icelandic landscape, with its volcanoes, geysers, earthquakes and avalanches, is incredibly dynamic, so too are images of landscape in Nói Albínói often in motion: the changing ViewMaster scenes, movable jigsaw puzzle pieces, palm tree decorations on a serving of birthday cake. While the landscape moves, Nói does not; he is unable to achieve physical distance from a landscape he can only see as oppressive.

By means of this unconventional, intentionally awkward, and psycho-perceptually ambivalent space, Pétursson calls attention to the act of viewing and experiencing landscape. Likewise, by evoking in the viewer an uncomfortable bodily relationship with the projected light that is the physical quality of a viewed film, he offers a case study for reassessing what it means to view and physically relate to art and images, just as we watch Nói's ambivalence between his physical and psychological relationships with the natural world around him. Given the deleterious consequences of our actual ambivalences towards and disconnection from the natural environment, it is crucial to investigate how contemporary visual culture addresses this issue—and how it might be used to reshape an understanding of landscape and an idea of nature that is not predicated on our distance from it.

Discussion

At a recent visit to an art museum in Reykjavík, I encountered a number of contemporary sculptural works whose mixed media, as indicated on the wall labels, were identified in English merely as "relief". As an art historian who believes in beginning a study of an artefact with the object itself rather than with theory, history or other frameworks, I found myself initially frustrated. I wanted to know exactly what these sculptures were made of when the materials were not self-evident, in order to understand their materiality and their artists' methods of production. But upon reflection, I began to question whether "starting with the object" in contemporary art sometimes has less to do with the material of fabrication and more to do with the material, corporeal effect on the viewer—and even the art historian, who herself is not immune to the experiences of desire and a bodily reaction to her objects of study. In other words, rather than having to do with the object in and of itself, materiality might be reinterpreted as an object's physical relationship with, or palpable effect on, the viewer.

In Dagur Kári Pétursson's 2003 film 'Nói Albínói', however, the viewer's spatial and perceptual relation to the image is ambiguous and ambivalent, as is the relationship of the protagonist to his physical surroundings. In studying film—a medium that, when viewed, is immaterial light—to talk about the viewer's physical relationship with the work is to speak more of his or her imagined interaction with the world within the film, as if it were three-dimensional. Yet it also concerns the psychoperceptual effect of the light projected onto the viewing screen and the sound of the dialogue and/or musical score. Pétursson's film not only complicates the viewer's perception of the space he has created, but also shows within his film a character whose sense of space, place, and landscape is distorted, leaving him misplaced and struggling to find a physical distance from the environmental source of his psychological distance.

The narrative tells of teenage Nói's (Tómas Lemarquis) alienation and fantasies of escape from his uninspired life in the remote and frozen Icelandic Westfjords. Yet it also sets up a tension between intimate and remote space, both diegetically (within the film) and for the viewer; space, and specifically landscape, takes on its own narrative and rhetorical roles. From the viewer's perspective, the space that Pétursson has created is often claustrophobic and stifling in its cinematography, but perceptually distant with the film's idiosyncratic use of diffuse light and mostly desaturated, cool colours. This echoes the tension that Nói feels between his desire to escape to a faraway, tropical Hawaiian landscape and his perception of that landscape as something very close and personal: a handheld image seen through his ViewMaster toy. In addition to visual referents of tropical climes that surround Nói even as he imagines he might find relief in their geographical distance, the movie's narrative is punctuated with shots of Nói's tiny hometown nestled against a snow-covered mountain that dominates the picture plane and forms a looming presence throughout the film.[2] The contrast between the numerous warm or tropical referents and the actual setting of the story furthers that tension that underscores Nói's alienation. It emphasizes his feeling of being out of place, or mis-placed, and his sense of an insurmountable distance from where he wants to be, physically and psychologically.

Examining four scenes from the film reveals this psychoperceptual push and pull between distant and intimate space in Nói Albínói, as well as the strain it creates *vis-à-vis* the viewer's "material" or bodily relationship with the moving image and the characters therein. Pétursson establishes this from the beginning in the title sequence of the film. Upon the introduction of two main characters, Nói and his grandmother Lína (Anna Friðriksdóttir), we see that they demonstrate opposing approaches to facing the snow that has enveloped their house, literally trapping them within the harsh Westfjord winter. While Nói sets to work shovelling out, his grandmother uses a bucket to bring the snow inside, and comically tries to melt it in the sink with tap water running as she walks away.

During this scene, multiple instruments in the score by Pétursson and co-composer Orri play the same melodic line but slightly out of sync. The double line, mostly of guitar, accentuates the disjunction between Nói and his grandmother; this disengaged relationship between Nói and other characters or settings heightens as the narrative unfolds and creates possibly an empathic disjunction between the viewer and the protagonist. Indeed, the music throughout the film contributes to its spatial ambiguity, which is what Pétursson and Orri intended in composing a score that is meant to evoke a feeling of warmth: music that doesn't fit with the

---

[2] 'Nói Albínói' was filmed in the Westfjord towns of Ísafjörður and neighbouring Bolungarvík; the mountain above Bolungarvík, Traðarhyrna, is the one we repeatedly see as a rhetorical, symbolic, and narrative element.

landscape.[3] Symbolically, one could say that the music's timbre is more fitting with the land to which Nói wishes to escape, the balmy beaches of Hawaii. Yet even though we feel the warm resonance of the harmonium and, in other scenes, instruments such as the banjo or kalimba (thumb piano) that evoke more temperate climes, there is a tension even within the soundtrack. This opening track introduces the musical saw, which sounds haunting and otherworldly next to the familiar strains of electric guitar. Yet it also conjures the sound of the wind, a sound that reminds Pétursson of Iceland itself.[4] As a light dusting of snow blows across the paved road and the musical saw gives way to the subtle sound of the wind, one can't help but wonder where this road leads to; can any roads lead away from a distant island?

Music aside, this title sequence sets up Nói's isolated struggle against the physical and cultural environment—an environment and an existential condition that other characters seem to have passively internalized. Here we see the blue pallor that predominates the colour palette of the film: the blue is a desaturated one that, throughout the film, takes on hues from the light greyish-blue of a robin's egg to a bleached-out turquoise; at other times the pallor turns towards a low-luminance pistachio or ochre green. In a way, this could be seen a derivation of—or even a play on—the "realist" style in Technicolor films: the blue/brown colour dyad, which sometimes uses green or red selectively as an accent, is transformed in Nói Albínói into a washed-out blue paired with a brownish yellow-green, and a highly saturated red appearing as a rhetorical device.[5]

The bluish tinge in the film takes advantage of a well-known aspect of colour perception in humans: the farther an object from the eye, the more blue it appears, since less light is necessary to perceive shorter spectral wavelengths like blues and purples. Thus the Icelandic saying "Distance makes the mountains blue",[6] roughly equivalent to the English proverb "Absence makes the heart grow fonder", is literally true because of atmospheric perspective. Symbolically, the blue colour of the mountains is beautiful, but beyond reach; illusory and untenable upon approach. The

---

[3] Interview with Dagur Kári Pétursson (2004).

[4] Author's interview with Dagur Kári Pétursson, 30th December 2005, Reykjavík.

[5] The "Realist Style", or "Dark Variation", has been identified by film scholar John Kurten as one of the four main colour themes used in TP-4 Technicolor films (1932–35). These colour designs were scrupulously controlled by Natalie M. Kalmus, Color Director of the Technicolor Motion Picture Corporation, from 1923–48. See Branigan (2005: 14–17). Also see Kalmus (1935: 139–147).

[6] *Fjarlægðin gerir fjöllin blá og mennina mikla* (Distance makes the mountains blue and the people great).

blue pallor of Nói Albínói, then, plays a trick on the eye: on the psycho-cognitive level, the colour keeps the film at a perceptual distance and creates an empathic distance between the viewer and the subject. But the spaces created by Pétursson and cinematographer Rasmus Videbæk are often flat and shallow. Many shots of characters throughout the film are at close proximity against a cramped wall or other space, leading to an uncomfortable claustrophobic feeling rather than a sense of intimate closeness. In outdoor shots, though the relative scale of characters or objects in the foreground indicates their distance from buildings or landscape in the background, atmospheric perspective is absent as the bright bluish-white of the sky blends into the same bright bluish-white of the snow that blankets the mountain, houses and ground. Again, this tension for the viewer between distance and intimacy emphasizes Nói's relationship with his environment. This perceptual ambivalence both challenges diegetic absorption (or the ability to "lose oneself" in the narrative) and calls attention to the nature of the filmic medium, the nature of viewing itself.

However, in talking with the director about the nuances of meaning in the colour of the film, Pétursson says that for him, colour is less important than light.[7] This is a sentiment that other visual artists and art historians in Iceland have impressed upon me as well—that the quality and amount of light in the country has a tremendous effect on the nature of many Icelandic artists' work, because it has such a tremendous impact on the perception and psychology of everyone living on the island.

In the next scene I would like to discuss, we see again the spatial dissonance as described above, as well as a tacit comment about light itself. Nói and his new girlfriend Íris (Elín Hansdóttir) have broken into a natural history museum and discover an interactive wall map of the world, whose scattered, small red light bulbs glow when a button is pushed for the corresponding location. Nói asks Íris to push a button to illuminate Iceland, but they discover that no such button exists. Asking if they should run away, Íris tells Nói to press a button at random to determine their destination. Íris turns to Nói without a word but with a smile and a kiss when the bulb on the Hawaiian Islands glows red.

Given the significance of light in Iceland, it is of consequence that Íris and Nói find this map in a dark closet, turn on a switch that lights the map from above, and can't find a button to light up Iceland. When Íris covers Nói's eyes with her hands and asks him to press a button to choose the place of their escape, he remains in the dark; perhaps for this reason does

---

[7] Author's interview with Pétursson, *op. cit.*

he select a location that might be warm and tropical, but is ironically also a small, remote, volcanic island whose image in the film recalls that of Iceland itself. While the kiss that Nói and Íris share is one of the only instances in the film of physical warmth or intimacy between characters, it is not a kiss of passion, but one of mutual comforting. The two nestle their heads against one another, and we see them from behind as, silently, they gaze at the map in front of them; they are no longer actively engaging with one another, but rather with the fantasy that the red light on Hawaii represents. Despite the characters' physical contact, the mood the scene evokes is nevertheless one of sad loneliness. It is here that Nói, alone, begins his dream to escape from one distant place to another, a distance from distance.

Once again, we hear a tension in the instrumentation: the light waltz of the piano that initially conjures a sense of curiosity as Íris and Nói look at the map, but then the dark, sombre, grounding sound of the double bass that makes its presence known as the map's red light illuminates Hawaii, suggesting the ultimate gravity of the fantasy. The music continues and the piano's melody brightens as we move from the inside shot to the camera's slow and steady pan across the sea, the tempo increasing slightly, as if to suggest Nói's possible departure—but as the camera pans, it comes to rest again on the town of Bolungarvík nestled against the mountain; the song ends without a sense of resolution. In this short oceanscape shot we move from one unknown piece of land across the water back to where we started, emphasizing again the conflict between near and far for both Nói and the viewer.

The tone of the very next scene in the movie is quite a different one, as Nói and Lína celebrate Nói's birthday. Lína has made tangible the landscape of her grandson's new fantasy, in the form of a whimsical birthday cake made to look like an island with plastic palm trees sprouting from the yellow frosting sand. She cuts the cake in her slow, deliberate manner, and physically moves this "landscape" by handing Nói a piece of the cake on a plate. This is one of a number of visually synecdochal references to landscape in the film: a rhetorical trope wherein a part of a landscape (or image of, or representation of a landscape) is used intentionally to evoke the idea of an entire, actual landscape or place. Other examples are the palm trees on Nói's father's (Þröstur Leó Gunnarsson) Hawaiian shirts, maps found in multiple scenes, Lína's jigsaw puzzle that has a picture of a giant mountain, even subtle things like the greeting card with a landscape image on the school principal's desk. Landscapes and references to them are everywhere in the film, surrounding Nói, reminding him—and the viewer—of the tropical place to which Nói longs to escape, as well as the Icelandic fjord he longs to escape from.

Unquestionably the most important landscape reference is the one inside of Nói's birthday present, which he unwraps as his grandmother passes him a piece of cake: a bright red ViewMaster toy. The scene in the kitchen leads into one in which Nói sits in profile in his living room, and he becomes entranced with what he sees as he looks (and we too look) through the ViewMaster. Here, Nói himself is the one moving the landscape, in a way, by clicking the lever on the toy and changing the scenes inside. But as mentioned above, the image of Hawaii in the ViewMaster is ironically similar to the way Pétursson portrays Nói's part of Iceland: with the same colours, the same sense of remoteness, and the loneliness of no one on the beach. Our protagonist gazes at a faraway land, but it appears flat and is confined to the intimate interior of the ViewMaster. At the same time, the image of Nói in his living room is also flat and shallow, and the wallpaper behind him is riddled with palm trees. Perhaps this is a suggestion that there is no difference between the place where Nói feels trapped and the place of his fantasies; Nói's imagination of Hawaii as a faraway escape curiously renders him perceptually oblivious to the Hawaiian imagery already around him.

Though in the film we see a few times just two images through the ViewMaster (an indigenous Hawaiian man and a empty, tranquil, white-sandy beach where we see a blue sky and ocean beyond the palm trees), we know that if Nói continues to click the lever, he will cycle through the different scenes and return to the same image. Metaphorically, it's as though he were taking the longest possible drive in Iceland, the highway that circles the perimeter of the island, only to return to the same place; his escape routes in Iceland are limited, just as his fantasies and means of escape from Iceland are limited, and perhaps myopically misguided, too.

This fantasy world inside the ViewMaster is the only world where the land itself appeals to Nói, where the landscape is beautiful in his eyes. But when Nói interacts with the actual natural environment around him, these scenes reveal an almost active hostility towards it as he takes his anger out on the landscape, expressing his resentment at being stuck in the Westfjords that for him are desolate and confining. After being expelled from school, for instance, in an act of futility a frustrated Nói throws rocks at a rainbow over the ocean; after a disconcerting interaction with his father, he shoots at massive icicles hanging in front of a cave with a shotgun. But when he views landscape for pleasure, he sits statically and gazes at the Hawaiian landscape as it passes through his ViewMaster; this intensifies the friction between Nói's desire to move towards that land but his inability to do so.

Coincidentally, Nói Albínói was released in the very same year as Peter De Bolla's *Education of the Eye* and Michel Conan's edited volume *Landscape Design and the Experience of Motion*, were published. In both publications the scholars launch an argument for the experience of landscape as an embodied and peripatetic one.[8] Their work critically calls into question the prevalent Western conceptualization of landscape as something stable and at a remove from humans who simply look passively at land before them. For De Bolla and Conan, people move and change within the landscape, just as, I would add, the landscape moves and changes with us. Just as the actual Icelandic land in particular—with volcanoes, geysers, earthquakes and avalanches—is incredibly dynamic, so too are images or synecdoches of landscape in Nói Albínói often in motion: the changing of ViewMaster scenes, the mobility of the pieces of Lína's jigsaw puzzle showing a mountain by the sea, the cutting and serving of a birthday cake in the likeness of a tropical island. But as we have seen, while the landscape moves, Nói does not; he is unable to achieve enough distance necessary from his own surroundings to see the landscape around him as anything but oppressive.

And yet, sometimes it is Nói who confines himself. In the final scene I will discuss, Nói's work on his new job digging graves for the local priest proves to be impossible because of the frozen graveyard ground. Frustrated, Nói throws down his pickaxe and walks to a phone booth, from which he calls Íris. When he tells her he can't come see her because he is on his way instead to see Gylfi (Kjartan Bjargmundsson), the fortune teller, Íris says with mild disdain that all fortune tellers say what everyone wants to hear: "They see changes ahead … unexpected money, a journey to exotic places…and a new person bringing love and happiness". Nói is optimistic that this can be his reality, but Íris tells him, "In your dreams".

Given Nói's misplaced optimism in this scene, the eerie mix of warm and cold in the music augments the narrative situation. The resonance of the harmonium, the kalimba recalling an African clime, and that windy, mournful sound of the musical saw evoke in their idiosyncratic combination a profound sense of sadness—a further sense of ambivalence of place and space. Nói, who fails to break the earth for the grave, simply walks away and is hopeful that the fortuneteller will bring him good news. But Íris remains sceptical, and during the phone call, Nói remains out of

[8] De Bolla (2003), Conan (2003).

reach in the confines of a claustrophobic phone booth. In other words, he is defeated by the land, thinks he can escape the situation, but ends up again trapping himself in a small space and put off even by his closest friend's resignation to undesirable circumstances. In a way, it is Nói's own fantasy that in this scene literally boxes him in and further disconnects him from the space and the people around him. As in other moments in the film, the viewer's separation too from Nói is augmented by a pane of glass—transparent, but not unnoticeable—as he is filmed from the outside of the phone booth.[9] The irony between the glass that mediates the viewer's spatial relationship with Nói and the close proximity of the camera to the protagonist again gives ambiguity to our sense of physical and interpersonal space, especially as it is in this closed-in box that he has one of his most poignant conversations. Yet his communication with Íris on the telephone is at a remove, and he finally chooses to visit the fortuneteller instead of her. And his act of closing himself off from the land around him, the land that thwarts him, is only self-defeating; though he is in a transparent glass box, his attempt is not to see through it, but to think only of the landscapes in his mind—to remain in a fantasy that ultimately will be all that is left for him.

Conclusion

In an interview, Pétursson describes 'Nói Albínói' as purposefully ambiguous, isolated, and in its own world.[10] When Nói retreats at several times in the film into his underground hideaway in his home, Nói himself is in his own world. Secluded, with no outside light, no view through any window of the natural landscape, in a space of his own construction and with no connection to the world around him, the incongruous warm light makes Nói appear at his most human—which might afford the viewer moments of psychological closeness with, or at least increased sympathy for, this character who otherwise remains at a remove. Yet Nói does not know how to act or what to do here. Whatever light or space he is in, the coldness, blueness, and remoteness of his natural environment remain inside of him and embody the profundity of his alienation. Though this is an existential, melancholy interiority with which the viewer might

---

[9] Other moments that involve characters being filmed through panes of glass include, for example, shots of Nói looking outside through windows, Nói and Gylfi shot from outside a window as Gylfi prepares to tell Nói's fortune, and Íris's father (Hjalti Rögnvaldsson) looking suspiciously out of his window when he sees Nói driving a stolen car.

[10] Interview with Pétursson, *op. cit*.

resonate, it is by its very nature something the viewer cannot touch. As Georg Lukács has written, "Nature…acquires the meaning of what has grown organically, what was not created by man, in contrast to the artificial structures of human civilization. But, at the same time, it can be understood as that aspect of human inwardness which has remained natural, or at least tends or longs to become natural once more".[11] Nói's internal landscape, his own nature, has been shaped by the landscape around him; try as he may to run from it, it is inescapable. He cannot achieve his distance from distance, nor can he reconcile his perceptual experience and his psychological experience. This speaks to the paradox that 'Nói Albínói' reveals: distance and closeness can be one and the same.

Being as it is so difficult to enter into the insular world of this film, what is it, then, to "start with the art object", the object that is 'Nói Albínói', the object that is not an object at all but projected light? What is the materiality, or perceived materiality, of this film, and how does it affect the viewer corporeally? Between the directing, cinematography, and the musical score, the effect reflects for the actual viewer those experiences of Nói himself, of the viewer within the film: the deep feelings of ambivalence and dis-ease. As unsettling as it may be to feel a bodily instability—a push and pull between intimacy and remoteness—it is the very thing, I would argue, that makes 'Nói Albínói' so compelling a film. The narrative is interesting, but not profound; rather, what drives the film is the unique cinematic creation of space and the situatedness in space (again, for Nói and the viewer), insofar as they bolster that narrative and heighten its emotional cogency. As the tension regarding the issue of distance finally climaxes in the literal breaking point of the avalanche that takes the lives of all those closest to our protagonist, there still remains no spatial resolution. When Nói looks again into his ViewMaster, we know this is the only thing he has left: A hand-held fantasy, a space simultaneously so close and yet so unattainable. For the viewer, is this lack of resolution of a perceived bodily relationship frustrating, as initially was my museum visit in which I couldn't identify the materials of the "relief" sculptures?

Perhaps it is. And in that, Pétursson has succeeded. His film speaks to a contemporary ambivalence towards the natural environment that is not only evident in Iceland but in every human-populated corner of the earth. While Pétursson might have intended to realise a story of adolescent alienation that could perhaps be universally understood, he more

[11] Lukács (1972: 136).

powerfully evokes in personal and bodily terms a physical uncertainty about how we sense, and make sense of, the environment and our relationship to it as societies and as individuals. For those that care to consider it, we can reflect upon our globally injurious relationship with non-human nature in the way scholars like Conan and De Bolla might if they moved from their study of landscape aesthetics and perception to conservational concern: Is ours a relationship in which we look at landscape as a distant and fixed phenomenon, in which we have irrevocably disrupted and continue to distress the environment around us? Or is it a two-way street, where we are beings in and of nature, moving through and acting within it, feeling it move through and acting within us as well? Perhaps a different perception of our material, spatial, and causal relationship with the environment (just as with art objects) could change our descriptive and conceptual language and our understanding of how it affects us, cannot be removed or parsed out from us. Until then, despite whatever environmental policy reform and conservation laws that might come to pass, the natural environment and we too of and within it will continue to waver, like the avalanche beginning its tremble above Nói and the only physical landscape he knows, in ominous volatility.

References

Branigan, E. (2005) *Making It Colorful: Color Relations and Practices* Unpub. ms.

Conan, M. (2003) 'Introduction: Garden and Landscape Design, from Emotion to the Construction of Self.' In M. Conan (ed.) *Landscape Design and the Experience of Motion* Washington: Dumbarton Oaks Research Library and Collection.

De Bolla, P. (2003) *Education of the Eye: Painting, Landscape, and Architecture in Eighteenth-Century Britain*. Palo Alto: Stanford University Press.

Kalmus, N.M. (1935) 'Color Consciousness' In *Journal of the Society of Motion Picture Engineers* **25**(2): 139–147.

Lukács, G. (1972) *History and Class Consciousness* Cambridge: MIT Press.

Pétursson, D.K. (2004) *Nói (Nói Albínói), DVD* New York: Palm Pictures.

CHAPTER TWELVE

AFFECTIVE SPACE – LOOKING BACK[1]

SERGE TAMPALINI

This paper locates the concept of 'Sensi/able spaces' as part of the artistic and organising strategy of set design in the production of theatre performances. In the patios of the title of this volume, the environment is the theatre, the space is the stage, and the art resides in the construction of assemblages that attempt to disrupt received ways a seeing, in favour of more complex ocular processes that potentially offer not only alternative ways of seeing but alternative ways of responding to what we see.

Whenever I use the visual metaphor 'I see' to indicate my understanding of something, what I am ostensibly saying is that I have filtered through all the available data and selected only that which was essential to having arrived at my understanding. I realise I am being overly simplistic, but I do so in order to make the point that the impression of understanding is always conditional upon a point of view. The same filtering process, and the resultant impression, may be observed when we look more closely at the phenomenon of 'seeing' itself.

'Seeing' is a misleading assertion that implies the world can be understood merely by looking. The deception lies in how we respond to our visual field.[2] We respond as if we have 'taken in' everything we see, even when we know that our brain filters our visual field and selects only what is essential to accomplish a particular task. Thus, seeing is only ever a partial solution to our visual field. The significance of this insight lies in its implicit questioning of the impression of certainty associated with seeing.

---

[1] This paper owes much to the work of the philosophers Gilles Deleuze, Brian Massumi and Maurice Merleau-Ponty as well as the performance theorist, Peggy Phelan.

[2] Cognisant of the complexity of the term *visual field*, it is used here to indicate the view seen by our two eyes without any movement of the head.

Seeing has become so naturalised that it takes a conscious concerted effort to intervene into the habitual way we see, before we are aware of the problems associated with it. If 'seeing', as I allude to above, is only ever a selective practice, then strategies that intentionally disrupt the unquestioned assumptions and conventions associated with 'seeing' may serve as triggers for alternative ways of seeing. The compositional and organizational practice of affective space is one such strategy. In order to begin our understanding of affective space we must first rethink the concept of space.

> Space is not the setting (real or logical) in which things are arranged, but the means whereby the positing of things becomes possible. This means that instead of imagining it as a sort of ether in which all things float, or conceiving it abstractly as a characteristic that they have in common, we must think of it as the universal power enabling them to be connected. [...] either I do not reflect, but live among things and vaguely regard space at one moment as the setting for things, [...] -or else I do reflect: I catch space at its source [...] realizing then that [all things] live only through the medium of a subject who describes and sustains them; and pass from spatialized to spatializing space. [3]

Spatializing space implies the inclusion of time in our participation with space, time as duration rather than instantaneous.

As a way of understanding what I mean by affective space, consider what happens in those 'unguarded moments' -those brief transitory moments in which we are not conscious of being conscious -our immediate surroundings appear strange and unfamiliar to us. For an instant our normally stable world collapses. Sensory data about our reality becomes at best unreliable and at worst alienating. What is remarkable is how easily this feeling of uncertainty may be triggered and by such seemingly mundane events -like when trying to open a door, we 'push' instead of 'pull' or when the inverted world reflected in pools of water, temporarily suspends the sovereignty of our ordered world. In the unfamiliar sensory states of these 'unguarded moments', when the 'flipside' of reality, as it were, penetrates our instinctual defences, our customary way of inhabiting consciousness seems no longer reliable. This phenomenon is not dissimilar to the experience of aspect perception - where the viewer first sees something as one thing and then suddenly sees it as another.

As our new 'reality' refuses to yield any immediate sense, we are forced to re-engage our once familiar world in the same way as when we

---

[3] Merleau-Ponty (1974: 243-244).

encounter something for the first time. Ordinarily all the things that we allow into consciousness have passed through a rationalising filter, conditioned by what we have learnt. In these 'unguarded moments' when they are not, we are startled. The point that I am leading to is that we may be similarly startled when we find objects in our visual field without having been consciously aware of their presence. Once aware of them, the objects appear as if they have been there all the time waiting for us to see them. A phenomenological perspective reveals that: "The world is there before any possible analysis of mine..." [4]

Now consciously looking at the objects we find, perhaps because of their familiarity made strange by their unexpected appearance, that we experience them as 'looking back' at us. It is this characteristic of 'looking back' that I want to posit as the energising ingredient of 'affective space'.

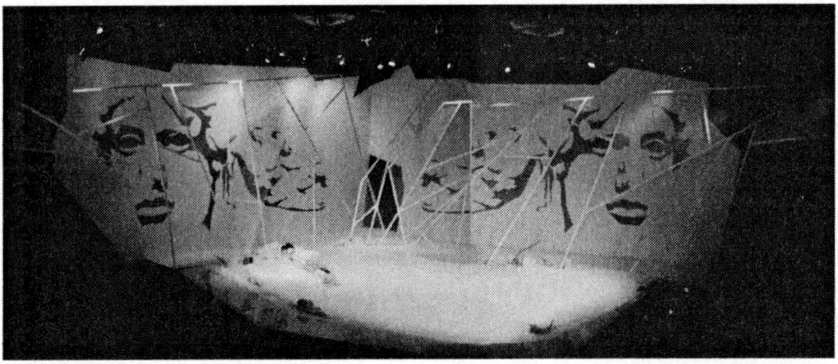

Figure 27: Louise Page: Salonika. [1985] Director: Raymond Omodei. *Design*: Serge Tampalini.

Traditionally, sight is understood in terms of vision. When I use the term vision however, I mean that particular characteristic of sight which connects 'seeing' to 'thinking'; that rationalizing vision that is both indicative and resultant of the habitual way of seeing the world that was fostered by our carers when we were babies, and developed by us as children. A way of looking that defines a relationship between a subject [which does the looking] and object [which is being looked at].

Consider, as I have already mentioned, what happens in the moment when the object appears to suddenly look back at the subject? I'm sure we have all had the experience of being caught looking at someone, when

[4] Merleau-Ponty (1974: x).

suddenly we become aware that they are looking back at us. It's like having to negotiate a moment of our reality, which under 'normal' conditions would not present any problems, however faced with the predicament of being caught unaware, we are forced to quickly assimilate this other space, into our familiar subjective reality. This is the experiential nature of the affective.

A possible first step in conceptualizing this other space would be to try and detach what one sees from what one knows –the eye from the brain; to intervene in the rationalizing vision associated with habit. The compositional strategy of set design for theatre performances is a useful place to start. All theatre design may be said to be already inscribed; doors, tables, chairs, stage properties etc. are already a kind of inscription. We do not see a chair, without relating it to the idea of a chair. This type of inscription functions as part of what I call effective space, and is exploited by the practice of theatre design as decoration. In order to have an affective space we need to rethink the idea of inscription. The inscriptions of affective space are twice removed from their counterparts in reality. Their concerns are more than reason, meaning and function – affective space has what might be called an excessive condition; one that participates in the conventions of space and time, but that seems to be full of them -an excessive condition of time and space. It is not that affective space has lost all meaning but that it seems to be full of meaning. This, in a sense, is the geometry of excessive condition.

In affective space, we understand that we can no longer conceptualise experience in the same way that we did in the grided space of space/time. Once the space becomes affective, inscribed with an 'other' logic, one that is no longer compliant with the rationalizing vision of the brain, then reason becomes detached from vision. While we can still comprehend the space in terms of its function and structure, somehow reason has become detached from the effective condition of seeing.

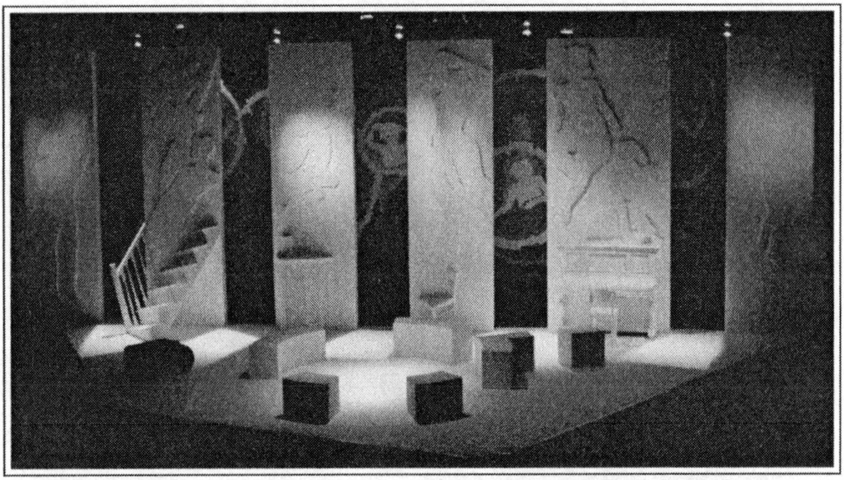

Figure 28: Ray Lawler: Summer of the Seventeenth Doll. [1988] Director: Raymond Omodei. *Design*: Serge Tampalini.

This detachment begins to produce a space that looks back; a space that gives the impression of having order even though it does not appear to mean anything. It does not petition to be understood in the traditional way, yet it possesses a logic that is the sense of something outside of our rationalising vision.

When Andy Warhol produced his canvases entirely filled with cans of Campbell soup, or with repetitions of the face of Marilyn Monroe, or silk-screened reproductions of newspaper photographs of car accidents, we were more interested in the point of view that wanted to do so, rather than the work itself. Looking back educes a similar response; aware of the space's excessive condition, we turn our attention to the consciousness that constructed it.

Effective space alludes to a literal connection between the object in reality and its representation. One way of inscribing affective space is to free aesthetics from the conventions of effective space. Abstraction is one method of doing so.[5] In my theatre work, (see figure 27–30) I try to push abstraction to a point where what remains is the least amount of informa-

---

[5] It would be foolish to try and conceive of abstraction as a homogenous concept; its construction and function varies significantly. For example, the abstraction of Wassily Kandinksy rejects classical figuration but tries to refine sensation by retaining abstract forms, but the abstraction of Jackson Pollock discards all "form" in favour of an active and chaotic texture of lines and colour.

tion required for the ontology of the work to be discernable. Reducing an idea to the utmost limit of recognition means that every smallest detail, every imperceptible difference becomes decisive. In the design for William Congrieve's play *The Way of the World*, I reduced the aesthetic and architectural considerations of the Restoration period to a point where what remained was just the suggestion of the playful wit and flamboyance of the era (see figure 29).

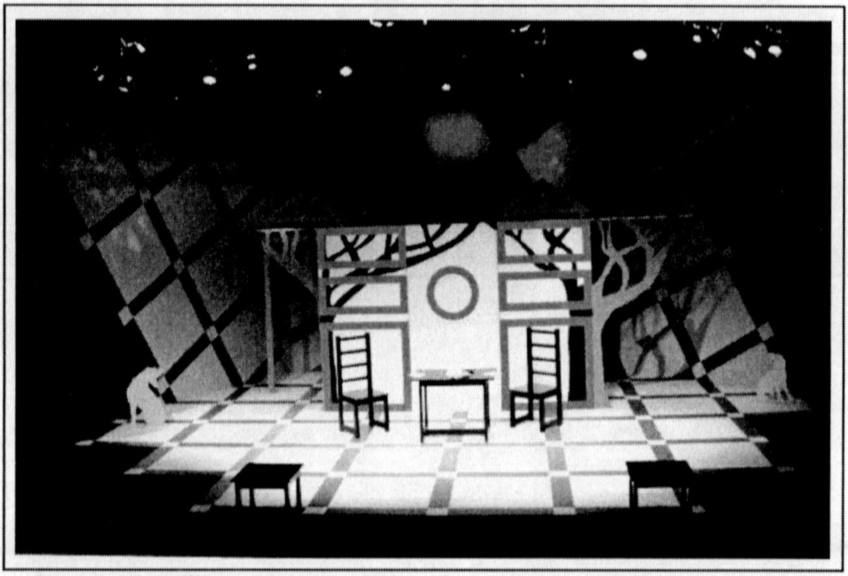

Figure 29: William Congrieve: Way of the World. [1986]. Director: Raymond Omodei. *Design*: Serge Tampalini.

Conventionally our view of reality is based on a visual language that configures the world within two frames: syntagmatic, i.e. the relation between one object and the objects that surround it, and paradigmatic, i.e. models of the world suggested by this relationship. In affective space models of the world and their basic syntagmatic structures have not been altered, there are just different paradigmatic selections that seem to disrupt established notions of how the world means.

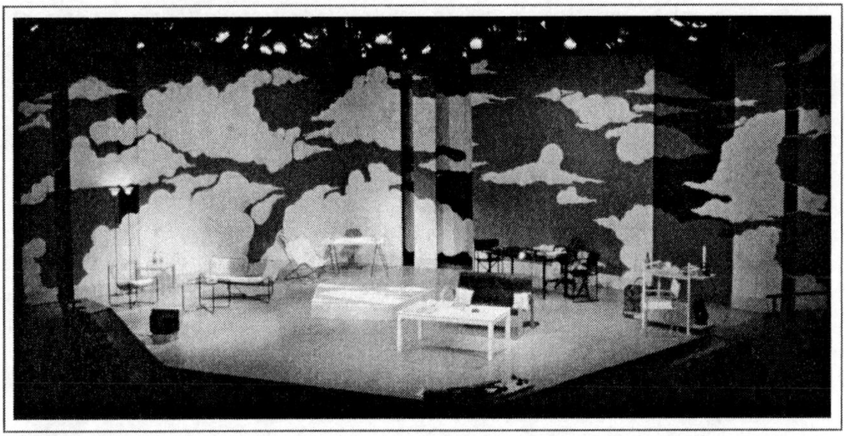

Figure 30: David Williamson: Emerald City. Director: John Milson. *Design*: Serge Tampalini.

The potency of affective space is not determined by its literal connection to conventional meaning. Its potency is discernible by its intensity. If there is a connection between the two it is one that bypasses the expectations of our rationalising vision and directly acts upon nervous system. It is in this way that intensity is a measure of the space's affectivity. It would be all too easy to fall into the trap of equating affective with emotional content, but arguably emotion is intensity owned and recognized...affect is not ownable or recognisable, and thus resistant to critique.[6]

The affective is never continuous or stable; it has an idiosyncratic mode of revelation, where it appears and disappears back from where it materialised, soliciting our attention, enticing us to enter undiscovered regions at the limits of our understanding and imagination. Its ontology is one of fluid geometry, intricate temporal modulations, layerings, and appearances and disappearances that give rise to an elaborate play between the observers received notions of consciousness and those released/offered by the space. It involves the willing suspension of the expectations allied with 'cause and effect' and linear temporality.

In concluding it is the usual suspects of liminality, ambiguity and paradox that ultimately emerge as the common devices/features of affective space. I use the terms 'ambiguity' and 'paradox' to denote moments where any precise meaning is uncertain; where signs exhibit seemingly contradictory characteristics. In so doing affective spaces not

---

[6] Merleau-Ponty (1974: 219).

only amplify and 'feed back' the embedded meaning that has gone into their construction, but also affectively multiply the audience's potential to make alternative/new meanings.

What I am finally suggesting is that the creation of affective spaces allows spectators the possibility of materialising something that exceeds their imagined modes of understanding and meaning making -of unravelling and making sense in ways they may not as yet have thought possible. As a theatre designer, I am as much interested in the audiences reaction to my design, as I am in its use by the director and actors…more interested in what something might become rather than what it is; more interested in holding open an object's potential to make different meanings, rather than closing its potentiality by concretising it into a form that is immediately recognisable and as such is relegated to the realm of the semantic, where it is reduced to a singular meaning.

References and bibliography

Berger, J. (1987) *Ways of Seeing* London: British Broadcasting Corporation and Penguin.

Bergson, H. [Trans. Mabelle L. Andison] (1965) *The Creative Mind* New Jersey: Adams and Co.

Gibson, J. (1974) *The Perception of the Visual World* Westport: Greenwood P.

Gregory, R. L. (1966) *Eye and Brain: The Psychology of Seeing* London: World University Library.

Huxley, A. (1959): *The Doors of Perception* Middlesex: Penguin Books.

Jay, M. (1994) *Downcast Eyes: The Denigration of Vision in the Twentieth-Century French Thought* Berkeley: University of California Press.

Lefebvre, H. [Trans. Donald Nicholson-Smith] (1991) *The Production of Space* Oxford: Basil Blackwell, Oxford.

Merleau-Ponty, M. [Trans. Colin Smith] (1974) *The Phenomenology of Perception* London: Routledge and Kegan Paul, London.

Pavis, P. (1982) *Languages of the Stage: Essays in the Semiology of Theatre* New York: PAJ.

Phelan, P. (1993) *Unmarked: The Politics of Performance* New York: Routledge.

Zeki, S. (1993) *A Vision of the Brain* London: Blackwell.

Zelanski, P. and Fisher, M.P. (1988) *The Art of Seeing* Sydney: Bay Books.

PART III

ENVIRONMENT

CHAPTER THIRTEEN

A MODEL SOCIETY
THE SKIN AND THE UNIVERSE WITHIN[1]

GÍSLI PÁLSSON

Focusing on the exploration of what molecular biologists refer to as the "universe within", this essay discusses the mappings and environmental/artistic manifestations of the new genetics. My take on this theme is partly framed within the context of the work 'A Model Society' by the artist Sarah Browne and the participants in her installation which involved photographs of models and woolen sweaters, lopi sweaters. I suggest we think of the iconic Icelandic sweater (lopapeysa) as being an extension of the skin of the Icelander's body. In such a perspective, the circular, coloured pattern typical for the lopi sweater—and, by extension, Browne's installation—hints at bodily signatures or birthmarks, in line with what many Icelanders, informed by modern genetic imagination and nationalist rhetoric, would refer to as a model genome, a relatively homogeneous genome ideally suited for analyses of human genetics and differential susceptibility to common diseases. The metaphor of the birthmark, I think, highlights some of the implications of the new genetics that has shifted the gaze of modern mapmakers from the immediate environment, the globe, and the universe to the cellular world of the body and its biomarkers.

---

[1] I am indebted to Sarah Browne for her extensive discussion of early drafts of this essay, for giving me access to detailed information about her work, and for the permission to reproduce the images relating to her project 'A Model Society'. Her highly informative and insightful comments significantly added to my understanding and the arguments presented. Also I acknowledge useful discussions of some of the arguments presented here with Edward H. Huijbens and Örn D. Jónsson. Needless to say, the responsibility for any errors there may be rests with me.

Wool Becomes Cool: The Icelandic Lopi Sweater

Sheep have been an important component of the Icelandic economy from the time of settlement in the ninth century. Although at times they have had devastating effects on vegetation through overgrazing of vulnerable commons and pastures, in the course of Icelandic history sheep and humans have generally fairly successfully adapted to each other and the northern climate—a setting more hospitable, though, than the name and the arctic location of the country may suggest, given the relatively mild winters and modest seasonal fluctuations in many respects, thanks to the Gulf Stream. Practically every part of the animal was somehow used for domestic consumption or export—meat, bones, intestines, skin, and wool. The wool is complex, with different colours and textures that make it useful for many kinds of clothing. For centuries, woollen cloth was an Icelandic trading staple. Knitting was introduced to Iceland in the Middle Ages, probably from England, and soon knitted woollens replaced the woven wool (vaðmál). The currently fashionable lopi sweater, however, is a fairly recent design. Although the lopi pattern is obviously decorative, it also has a utilitarian, practical function. Usually placed at the wrists and waist (bottom) of the sweater, as well as the yoke, the double or triple thicknesses of the wool at these points (created by the stranded method of knitting the pattern) serve to make it significantly thicker and warmer.

The exact origin of the 'typical' lopapeysa construct—a combination of particular wollen material, knitting style, and, last but not least, circular patterns of different colours—is a matter of some debate. In one theory, the characteristic circular patterns draw upon 'traditional' sweaters worn by Greenlandic women. Others suggest the patterns originated in Southern Sweden by the middle of the last century. Thus, the first 'true' knitted yoke sweater has been attributed to the designer Anna-Lisa Mannheim Lunns, born 1904, who worked for Bohub Stickning in Sweden.[2] Still others maintain the design is uniquely Icelandic. Many Icelanders like to believe it is a native thing and often they imagine it is much older than it really is. As McGregor points out, Iceland 'almost accidentally' became a knitting success story, and quite recently:

> First the circular yoke was applied in Iceland to the traditional (or, by 1950, not so traditional) local jerseys, using Norwegian and other patterns. This was a tremendously popular innovation in the 'fifties and 'sixties. Then an enterprising Icelander discovered she could knit with lopi which had not been thought of before as being suitable for knitting. It represents

[2] McGregor (1984: 131).

> only one stage in the spinning process; the wool has been carded and teased out for spinning but not yet spun.[3]

The success, however, of the knitting story undermines any claims of the authenticity of the products. Most 'Icelandic' jerseys "have only the remotest connection with Iceland, ... lopi is now as likely to have come from the back of British or Falkland Island sheep as from an Icelandic one".[4] This is even more true now than two decades ago.

The debate on the history and authenticity of the lopi sweater is interesting in its own right as it resonates with other debates (in both Iceland and many other contexts) about purity focusing on, among other things, nature, language, text, culture, and genome. Here as elsewhere, such debates reflect concerns with imagined communities, nation making, and representation. Whatever the history or origin of the design—an issue which is difficult, if not impossible, to settle once and for all independent of situated accounts and some form of representation—it seems the popularity of the lopi sweater was established soon after independence from Denmark in 1944, in the wake of the nationalist movement. While it has for long been popular among tourists, locally it fell out of fashion for years. Young people in particular saw it as an archaic, ethnic relic only fit for limited outdoor activities in bad weather. During the last years, however, it has gained a surprising boost embodying new kinds of significations, enmeshed in late-modern capitalism and postmodern playfulness. Not only is the lopapeysa both a working outfit in almost any kind of weather and a fashion thing for both men and women and for all generations, at times it functions as a kind of national costume also appropriate for public occasions. Currently, there is a lot of experimenting going on with material and design. Sometimes the typical lopi design appears on sweaters made in other material than wool. While such experimenting is exploring new avenues, it underlines the historicity of the lopapeysa concept itself.

Zooming In, Beyond the Skin

The notion of 'environment' usually refers to that which surrounds—from en 'in' + viron 'circuits' (from virer 'to turn'). An environment, then, includes just about everything, except that which is surrounded, a person, animal, or plant. To meaningfully operate in an environment, an organism must rely on some procedures of orienting or way-finding: embodied

---

[3] McGregor (1984: 106).
[4] *Ibid.*

mechanisms that adjust metabolism and movement to place and context, mental or cognitive maps that somehow decode or internalize the contours of the environment, or, alternatively, the construction and use of physical helpers or tools beyond the soma—some form of physical maps that inscribe such maps in material form or mnemonic device. With the advent of *homo viator* during the medieval age of exploration and discovery, mapping became a major industry and a pressing concern. To map was to colonize, to exercise territorial, 'orientalist' control. As Wilford points out, ancient maps were typically "oriented east (hence the expression 'orientation') presumably because that was the direction of sunrise".[5]

With the development of the new genetics during the second half of the twentieth century (the scientific discoveries and practices that enabled the visualizing of genetic material), mapping was increasingly oriented to the mapmakers themselves, to the minute contours of human genetic material. Academic textbooks nicely illustrate the language and imagery of the new genetics. Thus, standard textbook describe molecular genetics with the grand terms of discovery and voyaging:

> After many centuries, we have built up an approximate understanding of our external universe, but the universe within us has only very recently been the subject of serious study. The application of microscopy to the study of cells and subcellular structures provided one major route into this world, to be followed by pioneering advances in biochemistry and then molecular biology. Now, as we enter the next millennium, we are on the threshold of a truly momentous achievement that will have enormous implications for the future. For the first time, we will know our genetic endowment—the sequence of our DNA. Then our voyage into the universe within really will have begun.[6]

With their rapidly enhanced ability to zoom in on the minute details of hereditary material, geneticists and molecular biologists have established a powerful language of genomic cartographies. Several other kinds of explorations extend the biomedical gaze to the corporeal world; this includes computer images to guide surgery and brain function imaging or 'PET scans' (positron emission tomography) which reportedly represent the human brain at work in different modes. Figure 31 shows an advertisement for a recent technical innovation in surgery with the punch line "The Doctor Can See You Now". The advertisement partly owes its power to ambiguity in the relative emphasis on 'you' and 'now'; not only is the doctor available now, equipped with powerful machinery, she is, at last,

[5] Wilford (2000: 8).

[6] Strachan and Read (1999: 295).

capable of seeing the person involved. Or so the story goes. Thanks to the brain scanner, it is assumed, personal identities are finally in sight. In the current era of biomedical technology and new genetics, visual markers of one kind or another establish who you are—brain scans and genetic fingerprints.

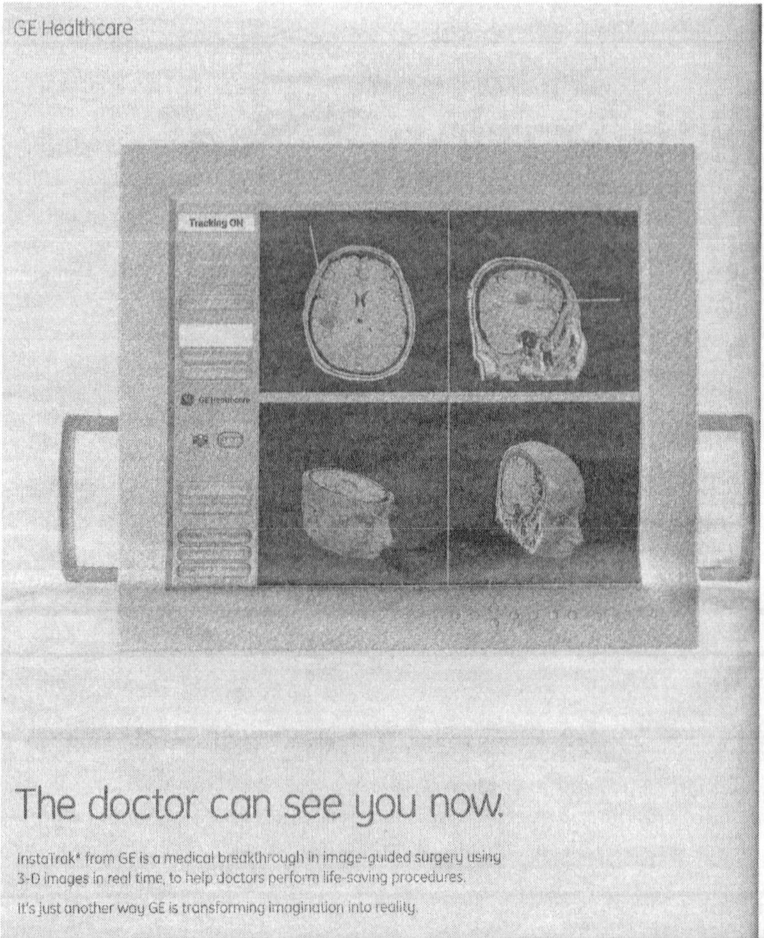

Figure 31: "The Doctor Can See You Now"
Source: GE the Financial Times.

Could it be that lopi sweaters with their circular patterns of different colours perform similar roles, classifying people and establishing personal identities? Perhaps they should be seen as remainders of roots, belonging, and the search for the past as well as a commentary on the hyper-modern world of genomic mapping in the age of the new genetics, on voyages beyond the skin into the "universe within". Interestingly, in earlier times folk knitting served to situate people in social space (not necessarily by intention, though, but by default), due the low frequency of travel and over smaller distances.[7] In some contexts, the use of particular patterns, colours, or dyes identified people of a region, village or even a single family or farm

'A Model Society'—A Model Genome

Sarah Browne, an artist based in Ireland, has organized a participatory project, 'A Model Society', that produces photographs and installations involving lopi sweaters (see figures 32 and 33) and publications (including an editioned artist's book and a forthcoming pattern-book). The project was undertaken as part of *Site-ations: Sense in Place*, year long site-related arts projects in 2005–2006 across the peripheries of Europe, involving an exchange of sixty artists from six countries: Wales, Latvia, Poland, Spain, Iceland, and Ireland.[8] In starting her project, Browne approached her participants in Reykjavík, anyone who "ever wanted to be a model", placing an ad in one of the major local newspapers (Fréttablaðið) in both Icelandic and English. Following initial contact, the respondents, males and females aged between 5 and 85, were asked to complete a survey. Prior to this, the survey was circulated to local personal and professional contacts. In all, over fifty people participated. Focusing on particular issues pertaining to social life in Iceland, the survey questions included "What are a) the best, and b) the worst things about living in Iceland" and "How would you describe Icelanders' relationship with the landscape?".[9]

---

[7] See McGregor (1984).

[8] See: www.site-ations.org/senseinplace (viewed: 1st February 2007).

[9] Browne (2005–2006).

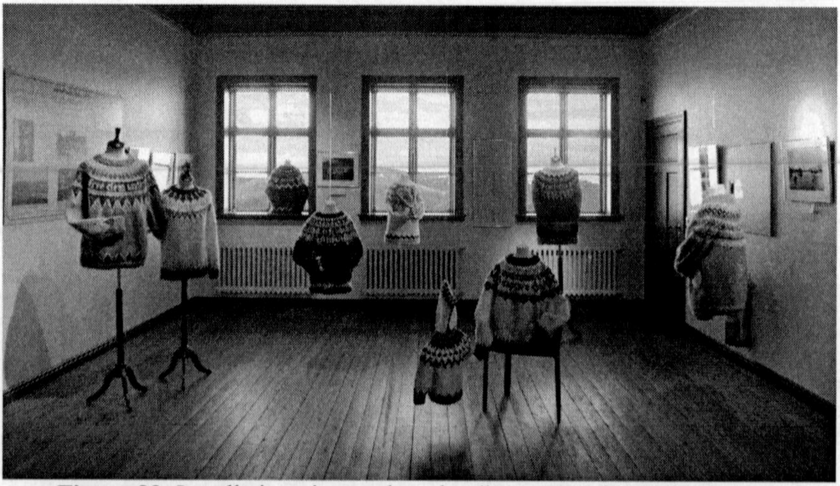

Figure 32: Installation view at the schoolhouse on Viðey island. Sweaters left to right: Free Clean Water; Rotten Politics; Gossip; Safe Jobs/ High Taxes; The Bank Owns Us (mittens), Awesome Daycare (seated), Exclusive/ Isolation; No War. *Source*: Sarah Browne: A Model Society, photograph, 2006.

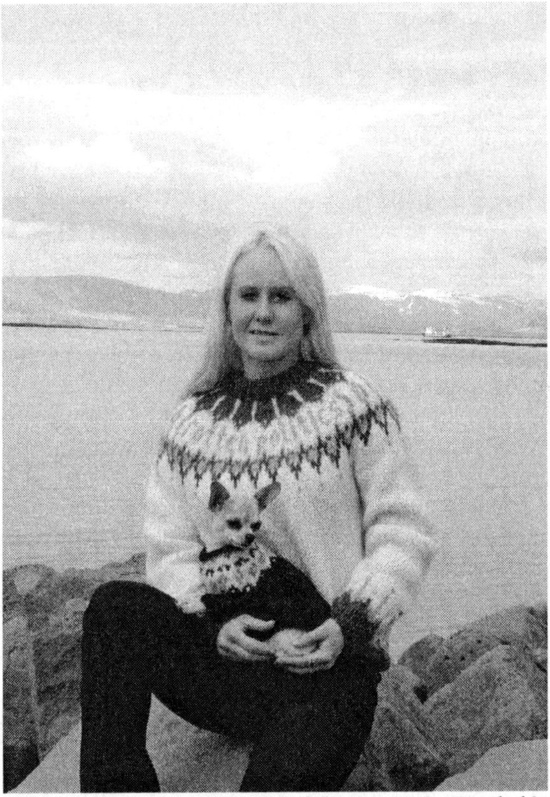

Figure 33: Rotten Politics. Model: Arna Bára Karlsdóttir.
Source: Sarah Browne: A Model Society, photograph, 2006.

Together, the respondents formed a kind of virtual community, maintained through correspondence via email and a series of photoshoots. Browne retained an editorial role throughout the project, selecting texts from the survey responses, knitting sweaters with the selected texts injected, and choosing models to wear them and pose for her camera. Each participant modelled a sweater that displayed the text he or she had contributed. A number of criteria were weighed up in order to represent a range of texts as well as a range of age groups, gender and so on. The selection also emphasized texts that were seen to be particularly expressive and interesting and that would physically fit on a sweater. While the knitting of the sweaters was a necessary step in the making of the work, the series of photographs really comprise the "finished" product, if there is such a thing in a durational and performative project of this kind. Such a project is necessarily complex, inviting multiple readings of what constitutes the "finished thing".

For Browne, the work is focused on idealized representations of the lopapeysa, in particular the romanticized visual language of knitting pattern photography that the work appropriates. These representations serve to sentimentalise landscape as well as interpersonal relationships, often promoting a myth of the nuclear family.[10] In the heyday of hippies and flower-power the lopapeysa seems to have signified something deep to Icelanders, especially young people, some kind of return to "nature" and simplicity.[11] The photographs that comprise 'A Model Society' are deliberately taken in such an idealized style, recapturing the nostalgia for the relationships with nature and society characterizing the 1960s and 70s, a time when the lopapeysa was becoming fashionable. This is one of the key points of the project. Significantly, one of Browne's participants (Bjarnheiður Jóhannesdóttir) referred to the Icelandic landscape as being "like the skin of the Icelanders' body: attached and the largest organ".

Another important feature of the work, in Browne's words, is "the participation of the models in the text eventually knitted into the sweater they eventually wear and pose in for the camera". Such an approach, Browne suggests, is in line with the "dialogical" model of contemporary art practice proposed by Grant Kester[12] in his book *Conversation Pieces: Community and Communication in Modern Art*. This seems to resonate with a research model that anthropologists sometimes identify as "collaborative ethnography",[13] in which subjects (the old-time "natives" and, occasionally, "niggers") become consultants, even co-authors, rather than just "informants". The aim of the 'A Model Society' project, Browne goes on, was "in some way to create a critical space in terms of how the lopapeysa is usually represented, as part of a wholesome, sentimentalised, even utopian identity", a space "opened up and articulated through the model's participation, interjecting texts about the society that were both positive and negative". The extracted texts, 13 in total, include "Long Workhours, Safe Jobs (front)/ High Taxes (back)", "Rotten Politics", "Free Clean Water", and "Exclusive Isolation". Browne reasons that the lopi sweater is "represented as the perfect expression of the particular Icelandic relationship between people and landscape, … typical of the national self-image". While the responses are particular to Iceland, she

[10] Browne (personal communication).
[11] Jónsson (2006).
[12] Kester (2004).
[13] Lassiter (2006).

suggests, "isolated from their context they have a general quality, as if speaking of a 'model society' ...".

On another level, 'A Model Society' can be read as a commentary on the notion of model genomes in contemporary biomedicine and the gene talk of many national contexts, including Iceland: the ways in which the signatures of the body (genomes with particular "ideal" characteristics) signal essences and identities of groups and individuals, conflating somehow the inside and the outside. Figure 34 shows a fragment of one of Browne's drawings, the means she used to calculate amount of stitches in her knitting (in this case, the extract Exclusive Isolation), how to place the text, what pattern repeat to choose, and so on. In some ways, it seems, her drawings relate to ideas of pattern and coding in the mapping of the body's contours and the sequencing of DNA, keeping in mind the reference to the "repeats" of patterns of four kinds of subunits—the chemical bases of A, G, C, T (adenine, guanine, cytosine, and thymine)—in discussions of the structure and change of hereditary material. The act of drawing, of outlining and arranging repeats, also draws attention to ruptures in established patterns, to genetic engineering, the deliberate alteration and design of genetic structures. Not only, then, does 'A Model Society' underline the conflation of inside and outside, it also underlines the conflation of nature and culture, emphasizing human fiddling with the birthmarks or biomarkers of the individual, managing "life itself".

In Browne's project, one may add, the notional community of models is taken as a kind of representative demographic sampling (0.0017% of the population, she reckons). The procedure of sampling and the statistical significance of the "findings", of course, are not the key issues, but rather the business of sampling itself, however crude or sophisticated. Once the notional micro-community has been established it is re-mapped back onto the larger population, much like in genomic sampling and mapping.

The exploration of the human genome is not simply driven by fascination with the unknown, with exotic landscapes and territories. It partly derives its momentum from the desire to understand diseases and find feasible, marketable cures, in particular "personalized medicine" geared to the genetic signatures of the individual concerned. Established in 1996, deCODE genetics is one of the important biotech companies on the scene. Based in Reykjavík it uses the Icelandic genome and the wealth of local historical records for the purpose of biomedical research and drug development. Drawing upon the presumed model genome of Icelanders, deCODE genetics has focused on several diseases. According to deCODE's early marketing strategy, the Icelandic genome is relatively homogeneous due to the geographical isolation of the population through the centuries and the fact that most Icelanders descend from a few settlers from the Viking Age. As a result, it was argued, it would be relatively easy to track down the presumed genetic causes of common diseases. It would help too that Icelanders had preserved detailed and complete medical records for decades and genealogical records almost back to the settlement period.

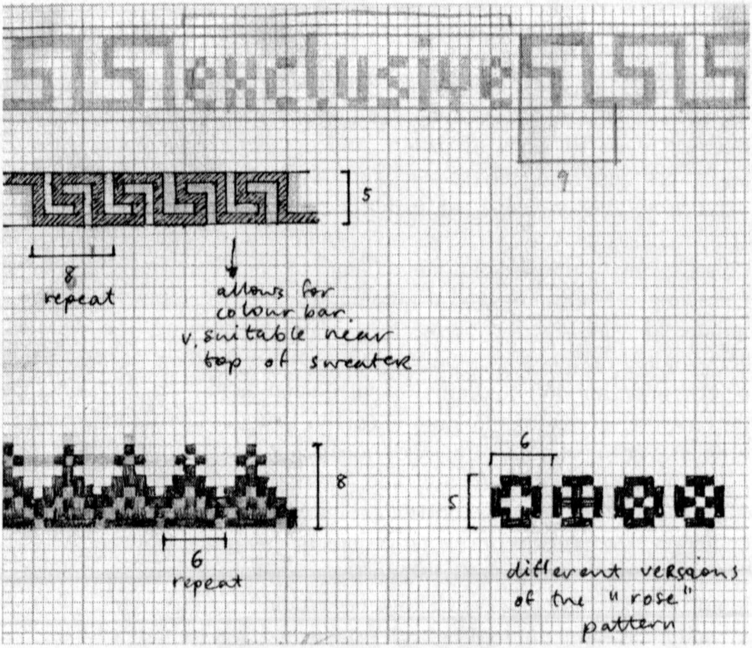

Figure 34: Exclusive Isolation
Source: Sarah Browne: A Model Society, details from a knitting pattern drawing on graph paper, 2006.

deCODE's early thesis of the homogeneity of the Icelandic genome has turned out to be a bit of an exaggeration although it does have an element of truth.[14] Through the ages Iceland has regularly been visited by travellers, in particular fishermen, and, likewise, Icelanders have frequently explored the larger world, to satisfy their curiosity or to look for work. Often genetic mixing has ensued. The North Atlantic Ocean does not simply isolate the communities it surrounds, it also connects—

[14] Helgason *et al*. (2003).

genomes as well as people and cultures. Also, in the current era of globalization Iceland is increasingly becoming a multicultural community. This is echoed in Browne's work. The participants in her project included people who are Finnish, Icelandic-American, Spanish-American, Spanish-American-Icelandic, from Bosnia Herzegovina and Sri Lanka.

While people from different races and cultures participated in the project, they were not "sampled" deliberately to achieve this effect. The racial/ethnic mixing that happened was purely unintentional and unanticipated, as Browne did not know people's country of origin, or in most cases even their appearance (before she met them some people did send unsolicited photographs). People's opinions (i.e. the survey and the text) were the chief criteria for selection. Perhaps the relatively high involvement of foreigners was partly due to the fact that the advertisement was placed in both English and Icelandic. The use of English text in the Icelandic context seems to have added something of an "internationalist" perspective. The effect of the multicultural sample, nevertheless serves to complicate and destabilize the discourse of "purity" or "authenticity" that pervades the model genome debate as well as debates surrounding "national heritage", or the "national signature" that the lopapeysa could potentially be. It may be noted that the troubled issue of "race"—one of the central issues in recent debates on research involving human genetic diversity (along with the issues of power, inequality, exploitation, and representation)—has always in one way or another focused on skin or "colour". As Haraway insists, in discussions of nature and roots race is always on the agenda, implicitly at least, representing a "fracturing trauma in the body politic of the nation".[15] Somehow the idea of "blondness" or light skin seems to loom in the notion of lopapeysa.

Not only is the thesis of the isolation and genetic homogeneity of Icelanders flawed, the claim that homogeneity would be ideal for genetic research needs qualification as much depends on the pragmatic context involved and the kind of disease studied. deCODE, nevertheless, boasts of several impressive successes, including the identification of a variant in a previously unknown candidate gene for type 2 diabetes and a recent clinical trial involving the risk of myocardial infarction. Results like these may have important practical implications, paving the way to diagnostic tests to identify people carrying the variant genes, preventive measures in terms of lifestyle and, possibly, the development of drugs that affect expressions of the proteins involved.

---

[15] Haraway (1996: 321).

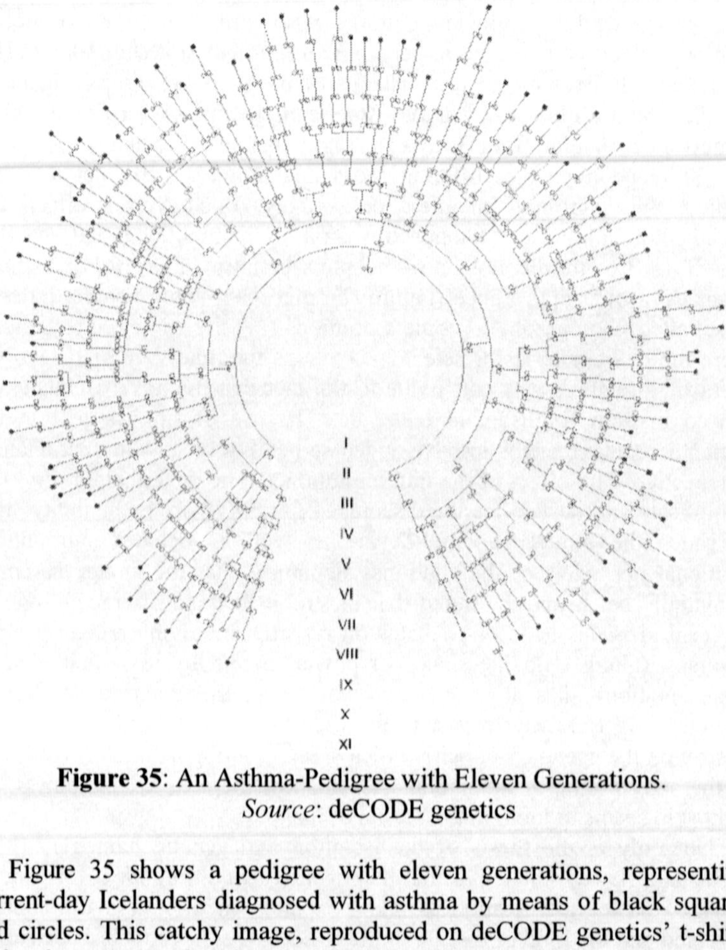

Figure 35: An Asthma-Pedigree with Eleven Generations.
Source: deCODE genetics

Figure 35 shows a pedigree with eleven generations, representing current-day Icelanders diagnosed with asthma by means of black squares and circles. This catchy image, reproduced on deCODE genetics' t-shirts and web pages, has functioned as the informal logo of the company. Apart from its visual appeal, the circular form may serve to foster a sense of unity and belonging, conveying an image of the encircling and gathering of the Icelandic flock, much like sheep are assembled in pounds (réttir) in the fall. The reference to the "circle" of the community and the continuity with the Norse past indirectly, at least, suggest a bounded genetic notion of citizenship, much like, perhaps, the lopi design. Wouldn't this genealogical "text", embodied in the second skin of a model, be appropriate on a lopi sweater?!

Dermography

For medical scholars, the skin has often been presented as the largest organ of the body. For social theorists, in contrast, the skin is an archaeological record with a series of temporal layers of social memory.[16] If the skin is packed with social memory, "dermography" (from the Greek "derma," skin) is ethnography, an embodied description of a way of life. In a sense, the skin serves as an (auto)biographical source of information, saturated with details about the body's identity. Such a description is not necessarily exact. Stacey suggests the skin diplomatically mediates fact and fiction; it is "the body's ambassador" in the sense that "it represents the interface between inside and outside …".[17] A huge cosmetic industry seeks to develop means for solving or hiding a whole range of "skin problems", rushes, acne, blackheads, etc., problems which often seem to run much deeper, in the immune system and the body's responses to itself, other bodies, and the rest of the environment. For some, focusing on the skin is to succumb to visualism. For others, however, the skin signifies connectivity among people. Bird-David argues that for the Nayaka of India the skin presents an opportunity to "stay in touch": "If in some Euro-American images the skin binds and isolates 'the body' from the 'outside'; in local experience equally at least it also, and equally, connects bodies".[18] With the powerful perspectives and tools of the new genetics our field of vision has been enormously expanded. As a result, I have argued elsewhere, a radical conflation of inside/outside has taken place:

> Not only has it become possible to zoom in on minute details, in the process the meaning of both phenotype and genotype have changed. Before the arrival of molecular biology, the scientific description and understanding of human differences was heavily reliant on phenotypic traits such as skin colour. The exploration of human differences, however, is no longer limited to visualization. At the same time, the underlying and hitherto invisible genotype is being rendered visible as a phenotype -- as a fingerprint or a birthmark.[19]

Birthmarks are some of the important signatures of the skin and most people have one or more birthmarks. Their cultural reading, however, varies from case to case. Australian Aborigines who call themselves *Yapa* or *Warlpiri* refer to birthmarks with a complex term, *kuruwarri*, which

[16] Ahmed and Stacey (2001), Connor (2004).
[17] Stacey (1997: 84).
[18] Bird-David (2004: 334).
[19] Pálsson (2007).

also denotes several other things, including ancestral essences and freckles. Biddle suggests that, to the Aborigines, through its constitutive force kuruwarri rejuvenates the species, controlling fertility, regulating social relations, and causing illness:

> kuruwaaru are embodied traces and imprints. Indeed, they provide a necessary material intercorporeal means for linking Ancestral bodies to Yapa bodies in crucial ways. It is not only in the country itself that Ancestral presence resides, but these presences … can enter women's wombs, cause conception and, in turn, leave birthmarks, freckles and other identifying traits of specific kinds of subjectivity upon individuated Warlpiri ...[20]

In some cultures, then, birthmarks represent a direct link to the past, the ancestors, the population, nation, or ethnic group, and the identity of the unique individual, much like, in some theories, the modern notion of genetic barcodes or fingerprints.

Conclusion

I have suggested that focusing on the Icelandic lopi sweater Browne's work 'A Model Society' can be seen as a commentary on the fascination of the modern era with model genomes and the ways in which they provide promising but complex avenues into ancestral connections and the mysteries of common diseases. Browne's project seems to address a multiplicity of viewpoints and identities that would complicate any idea of "model society" and the compromises that would be necessary to achieve such an ideal of constructed perfection. Given such an interpretation, the lopi sweater is the second skin of the Icelanders' body, a fragile membrane between self and society. By extension, its characteristic circular design is the equivalent of the biomarkers of hereditary material, underlining individual differences as well as common characteristics *vis-à-vis* other "populations". While all lopi sweaters are assumed to have something in common, a family resemblance or a collective birthmark in a metaphorical sense, no two lopi sweaters are completely identical, especially if they are handmade. The insertion of text in Browne's sweaters, the rupture in patterns, serves to make each sweater a unique combination, made to be worn by one particular individual. Thus, it underlines the idea of the individual birthmark, unique irregularities in the genome that, finally (with

[20] Biddle (2001: 180).

the new genetics), are both recognizable and subject to some form of human control.

In Icelandic, the notion of "stretching the lopi" metaphorically refers to an overly extended discussion, the excessive repetition or unnecessary elaboration of an argument, often in the negative sense as in "simply stretching the lopi" (bara að teygja lopann). At the risk of some "stretching", I conclude by saying that I hope my discussion has illuminated some aspects of the 'A Model Society' installation, drawing attention to the rhizomic threads of lopi beyond the surface appearance of orderly patterns as well as some of the complications of modeling genomes and societies. "Stretching the lopi", after all, may have quite important functions, in particular maintaining communion under the immanent threat of silence or misunderstanding. Much like, perhaps, the "junk" DNA of our genome.

References:

Ahmed, S. and Jackie S. (2001) 'Introduction: Dermographies' In S. Ahmed and J. Stacey (eds.) *Thinking Through the Skin* London: Routledge.

Biddle, J. (2001) 'Inscribing Identity: Skin as Country in the Central Desert' In S. Ahmed and J. Stacey (eds.) *Thinking Through the Skin* London: Routledge.

Bird-David, N. (2004) 'Illness-Images and Joined Beings: A Critical/ Nayaka Perspective on Intercorporeality' In *Social Anthropology* **12**(3): 325–339.

Browne, S. (2005-2006) 'A Model Society: Social Fabric' Unpublished survey.

Connor, S. (2004) *The Book of Skin* Ithaca: Cornell University Press.

Haraway, D. (1996) 'Universal Donors in a Vampire Culture: It's All in the Family: Biological Kinship Categories in the Twentieth-Century United States' In W. Cronon (ed.) *Uncommon Ground: Rethinking the Human Place in Nature* New York: W. W. Norton & Company.

Helgason, A., Nicholson, G., Stefánsson, K. and Donnelly, P. (2003) 'A Reassessment of Genetic Diversity in Icelanders: Strong Evidence from Multiple Loci Relative Homogeneity Caused by Genetic Drift' In *Annals of Human Genetics* **67**: 281–297.

Jónsson, Ö. D. (2006) 'Ull verður? Tilurð og fall íslensku lopapeysunnar' (Wool Becomes?: The Rise and Fall of the Icelandic *Lopi* Sweater). Unpublished paper.

Kester, G. (2004) *Conversation Pieces: Community and Communication in Modern Art* Berkeley: University of California Press.

Lassiter, L.E. (2006) *The Chicago Guide to Collaborative Ethnography* Chicago: The University of Chicago Press.

McGregor, S. (1984) *Traditional Scandinavian Knitting* Batsford: London.

Pálsson, G. 2007. *Anthropology and the New Genetics* Cambridge, UK: Cambridge University Press.

Stacey, J. 1997. *Teratologies: A Cultural Study of Cancer*. London: Routledge.

Strachan, T. and Andrew P.R. (1999) *Molecular Genetics. Second edition* Oxford: BIOS Scientific Publishers.

Wilford, J.N. (2000) *The Mapmakers* New York: Vintage Books.

CHAPTER FOURTEEN

THE USE OF LIGHT AS AN ENVIRONMENTAL TRACER

PIERRE AUBOIRON

This paper shows how, for a couple of decades, artists and architects have been using the universal metaphorical power of light to demonstrate the interdisciplinary nature of today's society by visualising environmental and societal features. Mainly based on the work of the French artist Yann Kersalé, this paper explores how light can embody phenomena, which may be invisible, or concepts, which are in essence invisible—such as time and human activity.

> *Tracer* (n): An identifiable substance, such as a dye or a radioactive isotope, that is introduced into a biological or mechanical system and can be followed through the course of a process, providing information on the pattern of events in the process or on the redistribution of the parts or elements involved.[1]

The use of light in urban planning and architecture has evolved into something much more complex than simply a tool that makes streets safer. Light has become a medium in its own right that has often been termed the fifth architectural dimension. We live totally immersed in a global visual system—the hegemony of the visual no longer needs to be proved—and two correlated results of this are: the hyper-mobilisation of our visual sense has been made to the detriment of other senses which now must be visually translated to be perceived, and the large-scale use of light as a medium may have become the only means to be visually "heard" by the majority of people in the ambient visual noise.

For a couple of decades artists and architects have been using the universal metaphorical power of light to demonstrate the interdisciplinary

---

[1] The American Heritage Dictionary of the English Language, Fourth Edition (2004).

nature of today's society by embodying and visualising environmental features. By using very complex lighting systems, made of captors and computers, they materialise and visualise both societal and natural phenomena such as noise, human activity, draughts and the current of a river or tide level.

This new approach to light is intended to make buildings fit back into their historical and socio-geographical environment and provide the public with a more accurate vision of the city and its main districts and buildings. Artists and architects have based their use of light on a social or environmental indexation in order to make concepts and features of everyday life visible and more tangible. Thereby they try and fight against the reduction of interpersonal communication in today's urban life which results from increasing visuality.

Artists and architects materialise phenomena we cannot perceive anymore because we have developed our visual sense to the detriment of our other senses. This could be compared to the use of tracers in biology, which reveal invisible processes or organisms. They do not create anything new; they just make visible what was already there.

Indexed Light

One of the most vivid modern metaphors of light is its allegoric embodiment of electricity. Although invisible, electricity has always been represented by bright blue sparks and flashes. In the collective consciousness light is the true substance of electricity. Therefore, it is not surprising to find on most appliances a light to indicate if it is in working order, or to put it in another way, whether it is powered by electricity. On a computer, flickering lights have always visually translated the complex data processing which occurs inside the hard drive. These lights are vital sensors and permit us to know that our ever more humanised computers are working.[2] Many are familiar with the image of Hal, the computer which held one of the leading roles in Stanley Kubrick's film 2001, a 'Space Odyssey'. Hal's physical presence was only manifest by its visual sensor: a simple lens lit by an inner reddish glow. It is worth noticing that Arthur C. Clark describes Hal as a simple 'spherical lens' in his epic. The red glow is Kubrick's addition; it allowed him to animate Hal with an

[2] Everybody has at some time anxiously stared at their computer lights waiting for them to flicker while hoping that hours of work had not been lost.

inner fire,[3] giving Hal a disconcerting human touch. This is directly linked with both the metaphorical and metaphysical aspects of light, the origin of which is lost in the mists of time. Since the origin of humankind, light has represented and embodied what is invisible and intangible as well as what has disappeared.

This complex metaphorical power of light may explain why architects and town planners have always obsessively sought an ever elusive mastery of light. The discovery of electricity and its large-scale generation gave them the first true opportunity to integrate light, as a material, in their projects. From this perspective, light which was initially used as a security aid, to make our streets safer, has quickly become a powerful tool to think, rationalise and signpost the City at nightfall: at night, a city is first announced from the distance to an approaching traveller by its diffuse lights in the sky. However, due to both the development and democratisation of new technologies, urban lighting schemes have entered a new age and with that an alternative and oneiric approach to light has emerged. This has lead to a significant break with the traditional way to comprehend the question of light in the city.

A number of artists have particularly emerged from this new approach to urban lighting: the Japanese architect Toyo Ito in collaboration with the engineer Kaoru Mende, the American artist James Turrell, the French light designer Yann Kersalé, the British architect Jonathan Speirs and the Swiss artist Daniel Schlaepfer.

Natural phenomena visualisation

Toyo Ito's Tower of Winds in Yokohama (1986) was governed by the desire to make this piece of architecture blend naturally into its immediate, functional environment. This tower was made as a ventilation and water tank facility for a shopping mall buried beneath it. A series of perforated aluminium panels form an oval cylinder all around the concrete structure. A very complex lighting system, also based on the use of captors, allows the materialisation and visualisation of 'natural' phenomena such as ambient noise and the movement of wind about the tower. The outer structure of the tower can look either like a brightly floodlit envelope or like a translucent film animated by colourful waves. A critic described the tower as:

[3] In some ways, both the metaphors of life and electricity can be superimposed: the electric current goes through an appliance as life flows through the human body.

> [an] Audio-visual seismograph [displaying a variation in lighting] ranging from glowing sparks to flashing strobe light, from geometrically regular intervals to amorphous shifts and flows, from retarded slow-motion to stroboscopic excitement, with all kinds of transformations and variants in between.[4]

Ito's work is based on a stratified perception of urbanism which can roughly be summed up in a two stratum system: the stratum of the urban fabric comprised of buildings, roads, rivers, etc. and the phenomenological stratum comprised of human influx, draughts, thoughts and data flows. According to Toyo Ito:

> This project is a conversion of the invisible rhythm and colour (made tangible through light, sound, air, etc.) of the city of which bodies are subconsciously aware due to a variable pattern of light.[5]

Due to a lack of funds and maintenance, the Tower of Winds has only been partially illuminated in recent years.

A time tracer

When Jonathan Speirs was asked in 1996 to design the lighting of the technical tower of the Bridgewater Hall in Manchester, he decided to turn it into a Tower of Time. There are three different light indexations: the interior lighting changes according to the Zodiac cycle, while light:

> on the exterior reflects the time of year, starting with green for spring, and running through yellow, red and blue, denoting each subsequent season in a gradual wash of colour.[6]

Last but not least, lines of light tubing delineate the eight stories and indicate the day of the week. This complex abstract clock obviously echoes with ancient observatories like Stonehenge and the ancestral desire to adjust human activity to natural cycles. After his collaboration with Ito on the Tower of Winds, Kaoru Mende put it that:

> we are getting fewer opportunities to enjoy the sense of changing time. Part of the reason is that we have become accustomed to lighting environments which always stay the same.[7]

---

[4] Neumann (2002: 204).
[5] Berwick (1998).
[6] Neumann (2002: 214).

Mende and the French visual artist Yann Kersalé have both taken up the challenge of using light in a very subtle and sensitive way in the over-lit cities of today, and found a way of lighting buildings that is almost organic. By highlighting the human, technological and media activities, which coexist and interact within buildings, they demonstrate the temporality and life form of the buildings. As the French philosopher and architect Paul Virilio put it:

> C'est [donc bien]à partir du temps intensif de la lumière, comme de la gravitation universelle que l'on devrait désormais étudier la durée-étendue, ce temps extensive de la matière des objets et des lieux. « Centre » ou plus exactement milieu du temps, la promptitude absolue de l'émission, lumineuse (égale à celle de la force de gravité), est la limite de la réalité. [...] Yann Kersalé n'est [donc] pas un éclairagiste, ni même un illuministe (encore moins un illuminé) mais un maître du temps, de ce TEMPS-LUMIERE qui compose le jour, le jour du temps de l'art de voir.[8]

By exploring this time-light, Kersalé discloses the sometimes discreet appearance of many concepts within everyday life.

A conceptual tracer

Born in 1955, Kersalé has worked with light for over 20 years and collaborated consistently with the French architect Jean Nouvel. Kersalé uses light as a societal tracer, which can make buildings merge into their historical and socio-geographical environments. This is particularly evident in a piece of work entitled the Ville-fleuve (The River-City). Until the 1920's, the river Loire and the river Erdre were not the only watercourses running in the city of Nantes in Southern Brittany; many of theirs branches as well as the river Sèvre passed through it, creating a network of swamps and islets; for this reason Nantes was also called the Venice of the West. Nowadays, all these canals have been covered by

---

[7] Mende (2000).

[8] Virilio (1994: 105) Author's translation: "We should now base the study of the notion of time-expanse; i.e. the extensive time of the matter of objects and places, on the intensive time of light as well as on Newton's law of gravitation. The centre, or middle [the French word milieu means both middle and medium] of time, the absolute promptness of light emission (equal to the one of gravity) represents the very limit of reality. [...] Yann Kersalé is neither an electrician nor an illuminist (even less a fanatic [the French word illuminé can mean floodlit, radiant, inspired or refer to a fanatic]) but a master of time, of this TIME-LIGHT, which makes up the day, the day of the time of the art of seeing."

roads, squares and buildings and have totally vanished from people's memory. Thus, the island of the Gloriette and Feydeau's island have become nothing but district names.

In his constant search for place memory, Kersalé decided to make the water re-emerge at night making it seep from the walls. In 1991 the city of Nantes gave him the opportunity to develop his concept by entrusting him with the lighting of its cathedral.[9] In this project,[10] the artist broke with the tacit rule of gothic cathedral lighting according to which light should only outline the building's verticality. At night, the massive cathedral of Nantes was totally disintegrated by 1,900 blue spotlights. Their blue hue varied constantly according to data recorded by captors set in the bed of the river Loire. The effect of this variation may have evoked Monet's series of the Cathedral of Rouen, therefore it was more related to the idea of visual impression than to the idea of representation. The nocturnal aspect of the building evolved from day to day depending on the strength of both the river's current and the ocean's tide (figure 36). To describe his concept, Kersalé has coined the term of 'geo-poetics.'[11] He says: "It is well known that the shape of a city changes, faster than the heart of a mortal."[12]

Indeed, the city of Nantes flourished due to its location by the estuary of the river Loire which could be considered as the venous system of the city while the cathedral was its mediaeval heart. Light was not used to reveal what underlay the building of the cathedral, the function of which was not only religious but political and economical. This building as well as the whole city of Nantes were born of the river Loire and were dependent on it. The lighting variation emphasises the fact that rivers remain fickle systems, though their courses can be altered. Nowadays we are more sensitive to the variations of figures, surveys and graphs than to environmental features.

---

[9] The original project was larger and included thirteen different buildings throughout the city of Nantes.

[10] It was intended to be a permanent installation. It was removed in 1998 in order to allow restoration work on the *façade*, but it has never been reinstalled due to the lack of support from local authorities.

[11] Here Kersalé obviously plays on both the meaning and the etymology of the word '*poetics*' which comes from the Ancient Greek *ποιεν* (to make).

[12] "*La Forme de la ville change plus vite, on le sait que le Cœur d'un mortel*", Introductive words of Jean Gracq in his book about Nantes.

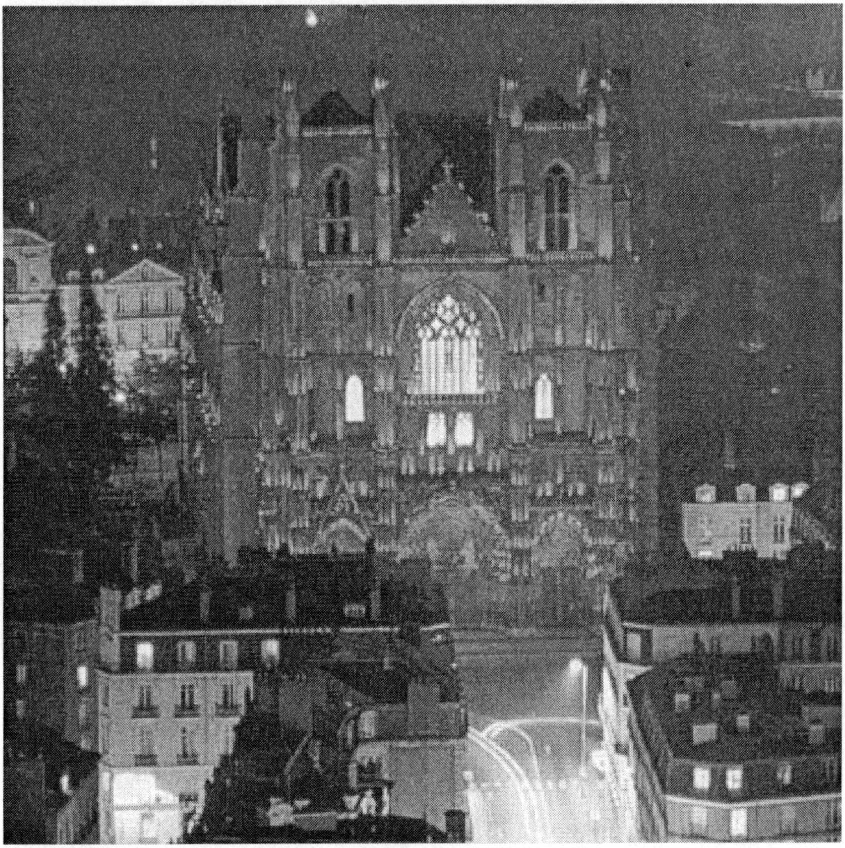

Figure 36: *La ville-fleuve* (1991-1998).
Source: © Yann Kersalé – AIK

Sensi/able Lights

A year later Kersalé worked in Lyon on another project called the Théâtre-temps (The Time-Theatre). This time he was asked to set up a new lighting system on the roof of Lyon's Opera house just renovated by the French architect Jean Nouvel. The red lighting of the roof changes depending on the level of "human activity" inside the building, the artist talks about "animal gestation" (figure 37).

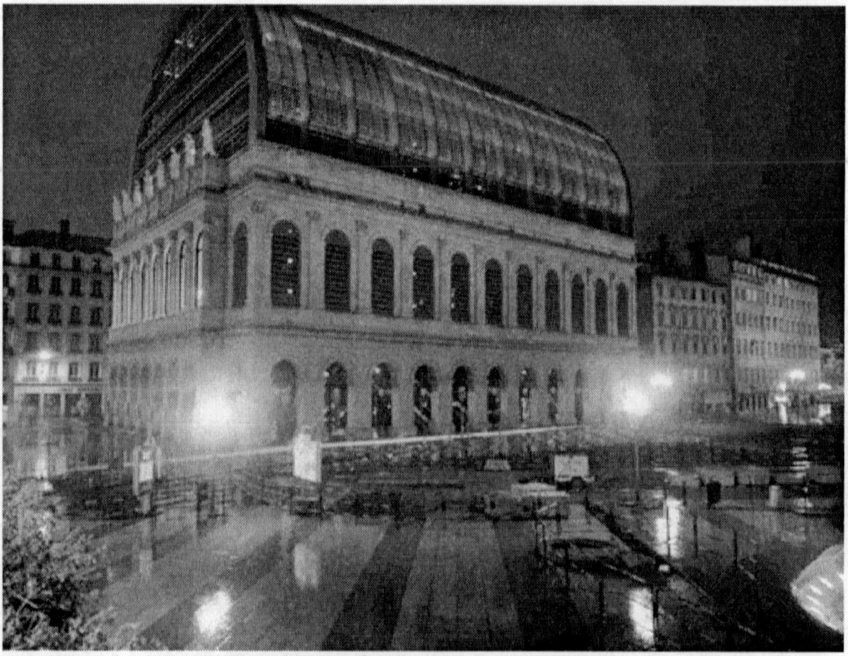

Figure 37: *Le théâtre-temps* (1993-).
Source: © Yann Kersalé – AIK

In the foyer of the building four screens show charts which translate graphically four vital activities of this living monster: costume makers and wardrobe masters in dressing rooms, technicians in the flies, actors on stage and the response of the audience. This activity is constantly recorded by captors and video cameras throughout the building. In this example the lighting puts the emphasis not only on a piece of architecture combining classical and contemporary styles, but also on the societal functions of the opera house which is at the heart of one of the main districts of the city. Kersalé tries to help us understand that a building has its own life form[13] and story to tell. His varying lighting tells us that an Opera house is constantly in operation, even outside opening hours. Thus a diffuse pulsing glow may betray overnight rehearsals or the night watchman can flare the glass dome up for a couple of seconds; and when there is no activity at all, only the first two lines of red neon tubes are kept lit and glow like live embers (figure 38).

[13] Light is unconsciously but deeply linked to the concept of life.

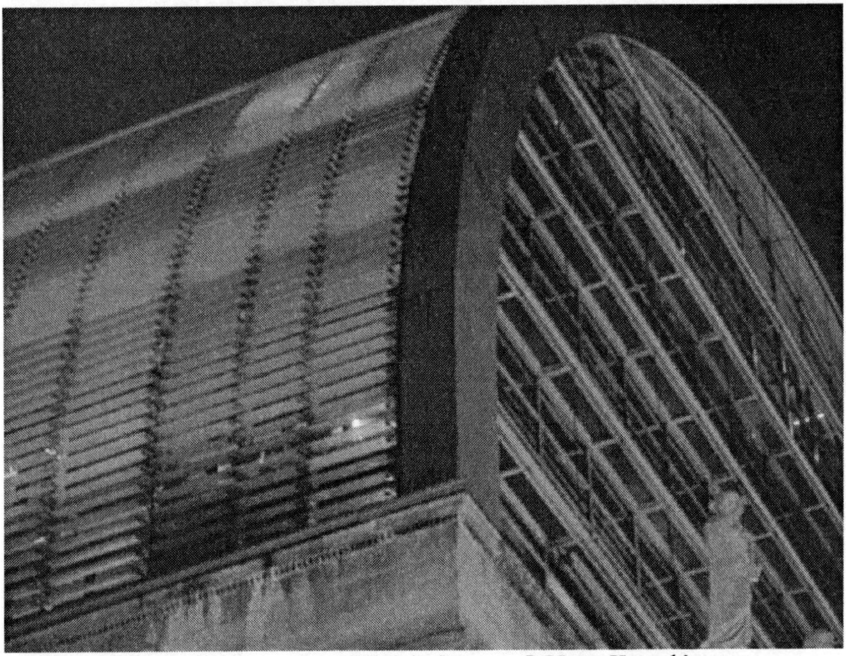

Figure 38: *Le théâtre-temps* (1993-). *Source*: © Yann Kersalé – AIK

In this project, light visually translates a flow of (human) activities and energies which goes unnoticed most of the time although crucial to keep intact the integrity of this microcosm, in other words, light pulses visually represent the vital inspiration of the Opera house as an artistic institution.

In 1997, James Turrell had a slightly different point of view while working on the lighting of the office building and computer centre for the natural-gas industry in Leipzig (Germany): the Verbundnetz AG building. Turrell decided not to translate the human activity occurring inside the building but the activity of the building itself. From this perspective, the building becomes sensi/able due to the ability of light to embody sensitive features and the very concept of life. This project echoes with one of the artist's suggestions as well as with light designers' concerns throughout the last third of the twentieth century: "Light should be a material with which we build".[14] The Verbundnetz AG building is totally self-sufficient in terms of energy due to both its own gas-fuelled power-station and to a system adjusting the heating and air-circulating systems. The artist

[14] Cooley (2001: 12, 14, 16).

decided to index his lighting to this autarkic technological world. The light colours vary according to the temperature that the energy supplier constantly reacts to and provides. Light is used as a physiological tracer to reveal thermodynamic exchanges between this artificial body and its outer environment.

The Swiss artist Daniel Schlaepfer plays with light in order to create sensible landscapes and to demonstrate the other side of our everyday visual environment. In 2003, he worked in collaboration with the architect Andreas Schmid on a new cinema in Lausanne (Switzerland). He designed a lighting system entitled 'Sensitive'. A very complex network of optic fibres, hidden within the double glass outer envelope of the cinema, is connected to sensory captors set in the pillars of the building. When one touches one of the pillars, the whole *façade* metamorphoses itself and seems to be shivering with light. Here, Schlaepfer suggests that everybody's action has got direct positive or negative impact on their immediate environment, even if, most of the time they stay unnoticed because they are invisible. This project and Turrell's one, both refer to the most recent cybernetic theories by using a certain form of communication between systems and their environment with a control (feedback) on the systems' function due to environmental changes.

Light as a haptic medium (conclusion)

In the nineties artists' and architects' main concern was to cast buildings in their immediate societal and environmental contexts in order to make them as meaningful as possible. New lighting technologies enabled them to make these concepts more legible and tangible in order to improve people's understanding and experience of architecture. From this perspective, they tend to offer a more organic and intimate perception of their buildings by depriving the latter of any precise outlines and invading them by light. Thus Toyo Ito explains that:

> the tower (of winds) with such physical presence loses its presence after sunset and metamorphoses into a phenomenon of light. I refer to this metamorphosis from an opaque substance into a transparent object of light as 'fictional'.[15]

A decade later, there is no choice but to notice there has been a paradigmatic shift in their approach to nocturnal reading of architecture: architects seem to assert the buildings' self-sufficiency. In their indexed

---

[15] Berwick (1998: --).

lighting, the recording of human activity does not seem more important than the activity of the air conditioning system; it is only one factor the monitoring of which enables the visualisation of the physiological organisation of a building.

Thanks to a sensible use of new technologies, they make concepts and aspects of our everyday life visible and more tangible. The strength and poetic quality of their work dwell on their ability to make us forget the technological complexity of their light installations for the benefit of the sole aesthetic experience. In this instance, computers, associated with light, perform like prostheses and compensate for our disability to comprehend our environment in its whole complexity. Here the artist's work does not deal with creating something new but with making existing things visible. His aim is not to produce any aesthetic[16] experience as the term is commonly understood, and he does not use light for its ability to mesmerise, but for its ability to embody the intangible and to express much more than simple visual sensations.

In some ways, this can be compared to the use of tracers in biology which reveal invisible processes or organisms.[17] Sight remains our last means to comprehend a world where we live totally immersed in a global visual cacophony from which it has become very difficult to isolate any particular message. Wind, air, feelings and ideas are totally fictional because they are not visible; in other words all that cannot be seen is barely considered anymore. Human visual sense has been hypertrophied so much over the last century that it now tends to supplant other senses. Thus, they teach us how to look again at our direct environment by looking beyond the static aesthetic coating[18] which now covers and conceals all aspects of society. They allow us to understand better societal and functional mechanisms which are hidden behind the *façades* of buildings. However some of these projects offer alternative sensory

---

[16] The term *aesthetic* has seen its meaning reduced by common *parlance* to an "evaluation process" only. This elides the crucial haptic aspect of such "aesthetic experiences".

[17] Here we must think of one of the most popularised scientific images, the electrophoresis of DNA under ultraviolet light. This very strong image is linked with the decoding of life. It is no accident that memory sticks or USB ports are increasingly lit with hues evoking ultraviolet rays.

[18] This means *Aesthetics* should not be considered as an ideal anymore. *Aesthetics* is now something between a societal norm and a self-inflicted torture. Nowadays, any object, from the most insignificant to the most astonishing, is always perfectly but impersonally designed. Public figures must appear larger than life in the media.

experiences to the viewer whose haptic sense[19] will be mobilised as well as their visual one. The haptic sense is synonymous with contiguity, contact and resonance, whereas the visual sense permits a transcendent, distant and, sometimes, disconnected point-of-view. For instance, it enables the perception of concepts such as weight, pressure, balance, temperature, vibration and presence.

One may object that these projects, often backed by city councils, are a way to improve the image of a district and to make people forget the social reality of such poor or industrial areas. Indeed, this is an emerging trend which contributes to a broader architectural and urban scheme. Building developers copy the above artistic concepts without trying to adapt them to the context of their new place of implementation. Here their purpose is not to bring back the memory of a place or to clarify the *raison d'être* of a building, but to entertain the passer-by. This has been very well summarised by the term *architainment*.[20] This term is a coinage of the architect Luis Fernández-Galiano and was born from the contraction of architecture and entertainment, and associates architecture with entertainment in luminous sceneries overloaded with special effects, video screens and direct references to Arcade games.

Be that as it may, all the artists mentioned earlier have based their work on a novel societal, environmental or functional indexation of light. Thereby, they try to fight against the self-isolation affecting more and more young people who find it difficult to find their place in today's urban life and society. Western societies are now revolving around the hypothetical concept of a single standardised fictional self. It heralds the end of any form of social cohesion and with this in mind specialists in the Humanities, such as Susan Buck-Morss, now agree that autism appears to be a societal interpretative model much more relevant than narcissism. New technologies have played a leading role in producing and maintaining this societal autism by disconnecting people from their sensible environment in favour of a so-called virtual society.

However, thanks to a sensible clever use of new technologies, artists such as Ito or Schlaepfer have allowed people to visually experience their haptic sense in the heart of the city itself where they were previously used to doing it through the flat screen of their home computer. By using the

---

[19] This includes tactile, kinaesthetic and proprioceptive senses, and is unique in that, unlike the passive senses of seeing and hearing, action (manipulation) always precedes perception. In the virtual world, the haptic sense is used to convey information in place of vision or hearing.

[20] Fernández-Galiano (2005).

metaphorical power of light to trace concepts as well as intangible aspects of our everyday life, they widen our perceptual field to non-visual sensations. In other words, they invite us to (re-)discover our immediate social and natural environment as well as our own senses.

References

The American Heritage Dictionary of the English Language, Fourth Edition (2004) Houghton Mifflin Company: Boston.

Berwick, C. (1998) 'The Tower of Winds and the Architecture of Sound, an interview with Toyo Ito' On the website of the magazine The Take: http://www.thetake.com/take05/take04/html/42ndst.html. Viewed on December 28th 2006.

Cooley, R. (2001) 'Interview with James Turrell: Adventures in Perception' In *Architectural Lighting* **16**(2): 12, 14, 16.

Fernández-Galiano, L. (2005) 'Spectacle and its Discontents Or, the Elusive Joys of Architainment' In W. Saunders (ed.) *Commodification and Spectacle in Architecture: a Harvard design magazine reader* Minneapolis: University of Minnesota Press.

Mende, K. (2000) 'Trends in Urban lighting in Japan' In *Philips Lamps and Gear Magazine* **4**(1).

Neumann, D. (2002) *Architectural of the Night, the illuminated building* Berlin: Prestel Verlag.

Virilio, P. (1994) 'L'instant lumière' In P. Curval, and Y. Kersalé (eds.) *Editions Monotypes* Paris: Hazan (103-105).

CHAPTER FIFTEEN

FIRST AND SECOND NATURE[1]

GIOVANNI TUZET

1

Pascal asked: Why not take culture as second nature and nature as first culture?[2] The same may be asked as to nature and artefacts: Why not take artefacts as second nature and nature as first artefact? Making abstraction from the metaphysical claims and the related ethical claims involved in such issues,[3] it is of interest to note that in contemporary poetry some authors have represented artefacts in natural terms while others have represented nature in artificial terms. So, there is no nature/artefact dichotomy in contemporary poetry and aesthetics. These are the two options worth discussing:

(i) representing artefacts in natural terms (e.g., the motor back, a river of electric fibres);
(ii) representing nature in artificial terms (e.g., the sun lamp).

The first consists of looking at new things with an old eye. The second, of looking at old things with a new eye. The first we may call *conservative representation*: it represents new things (artefacts) in old terms (natural terms). The second instead we may call *innovative representation*: it represents old things (nature) in new terms (artificial terms).

---

[1] The author thanks all the participants in the conference "The Nature of Spaces" (Reykjavik, June 2005) and, in particular the reviewers for this volume, who made very helpful comments on previous versions of this paper.

[2] *Pensées*, 120, ed. Chevalier. Cf. Montaigne, *Essais*, III, 10. It is, before Pascal and Montaigne, an Aristotelian theme (see book 2 of *Nicomachean Ethics*).

[3] Cf. McDowell (1996: 84-92); Bonito Oliva and Cantillo (2000). Recently, on the naturalization of reason see Papineau (2003); on the naturalization of meaning see Millikan (2004).

The question may be posed why the second option is an interesting one. I will claim it is such since, in general, it challenges our previous views. Then a further question may be posed, namely whether such an artefactual representation of nature is an interesting representation *of nature*. I will answer; it is, not only for aesthetic but also for ethical reasons, since, in particular, a representation of nature in artificial terms helps in seeing what is at stake in manipulating nature. (One might also wonder if the reasons I will offer make it even more interesting than a natural representation of nature). Before going into the argument, let me clarify that this paper is not on the appreciation of nature itself, but on the (appreciation of the) artistic representation of nature.

2

Conservative representations are largely used in Futurism's aesthetics. To represent modern life and artefacts, Filippo Tommaso Marinetti and others have recourse to natural elements, natural images and natural metaphors. For instance, in the Futurist *Manifesto* of 1909 (at point 4) a car is represented as a formidable creature, whose pipes are similar to "snakes with an explosive breath".[4] This kind of representation and style is quite peculiar to Marinetti and his followers. But other poets still belonging to the Futurist movement try the second way, namely representing nature in artificial terms. I will consider some examples taken from the works of Corrado Govoni and Luciano Folgore. Both these two poets belong to the Futurist Movement but, in my opinion, they are quite original in their interpretation of the Futurist spirit. While the orthodox Futurists want to represent the artefacts of modern life and tend to represent them in natural terms, such heterodox Futurists as Govoni and Folgore often represent nature and tend to do it in artificial terms. So, the orthodox Futurism generally uses conservative representations, while the heterodox Futurism uses innovative representations.

---

[4] Now in Marinetti (1983: 10). See also his poem *To the Racing Car* (*All'Automobile da corsa*) composed in 1908 and revised in 1921 (readable now in Sanguineti (1971: 547–549)): the car is a creature freed from the reins, monstrous and divine. For an English translation of some of his works, see Marinetti (2002).

3

Let me start from Corrado Govoni.[5] In his poem of 1911 *The Frogs* (*Le rane*), the singing of a nightingale is depicted as the coming out of a miraculous and infinite noodle of coloured paper from the bird's mouth, and the shrieking frogs are depicted as strange washerwomen washing in the lagoon the moon's wedding dress. In Govoni's poetry some Futurist suggestions go together with a peculiar sensitivity for natural elements and landscapes. In a poem of 1915, *Milan and Myself* (*Io e Milano*), the Milky Way is an immense washing of stars and the firmament an infinite ebony chandelier. In *Self-portrait* (*Autoritratto*), which is a calligram of 1915, the poet's nose is depicted as an elastic pump for odours and his lips are the 'pale sofa of the kisses'. A nose as a pump is a striking example indeed of how to represent a natural thing in artificial terms.

Now let me turn to Luciano Folgore. He provides some quite suggestive examples of the artefactual representation of nature, namely what I call innovative representation. In his poem of 1919 *World under Green Skylight* (*Mondo sotto lucernario verde*), he represents a wood as a city of trees and the sun as a green lamp. A wood as a city and the sun as a lamp: plainly these are representations of nature in artificial terms. In another poem of 1919, *Boundaries Made of Leaves* (*Limiti fatti di foglie*), he twists some natural and artificial elements, combining the landscape with human artefacts, so as to produce a complex scenario in which men and nature contribute to a surprising, miraculous richness.[6]

4

Of course, these issues are not confined to Italian poetry. Another impressive example of innovative representation may be found in Blaise Cendrars' poem of 1916 *La guerre au Luxembourg*.[7] Consider the following lines:

> A kite fast as a pursuit plane
> The trees go flat and the leaves fall and roll in a parachute
> The 3 veins of the flag swell at every shot of the wind howitzer.

---

[5] All the poems I refer to are readable now in Govoni (2000). For an introduction to Govoni's poetry, see Mengaldo (1978: 3–8).

[6] These two poems of Folgore can be read in Sanguineti (1971: 615–618).

[7] Cendrars (1967: 113) The translation used is mine. For an English translation of his poems, see Cendrars (1992).

A kite is represented as a pursuit plane, the falling of the leaves is represented as a parachute turn, the strength of the wind is the strength of a howitzer. Consider in particular the falling of the leaves as a parachute turn: a natural process is represented in artificial terms.

Other examples may be found. Similar combinations of landscape and human artefacts are present in Arthur Rimbaud's poetry, in his *Alchimie du verbe* as well as in his *Illuminations*. It can be claimed in sum that the innovative representation of nature is rooted in some poetry of the last part of the 19th century at least, and flourishes in the countries in which an avant-garde movement as Futurism flourishes.

5

Now, concerning space, it is possible to find a similar articulation: some artists put artefacts within a natural space, while others put nature within an artificial space. Compare a plane in the sky and a dog in a space station. The first is an artefact within a natural space. The second a natural entity within an artificial space. Such a difference can be found in contemporary art. Again we can take the example of Futurism. When Futurist painters as Fedele Azari, Gerardo Dottori, Fillia, Bruno Munari, elaborate in the 1920s and 1930s the so-called *Aeropainting*, they want to represent such artefacts as aeroplanes and the experience of flight.[8] They share the aims of what I call orthodox Futurism: representing the artefacts of modern life (even though the representation of the flight experience includes the different perception of the landscape). In contrast, when the Futurist painter Leonardo Dudreville realises in 1912-1913 his cycle *Seasons* he shares the aims of what I call heterodox Futurism: representing nature in a new way, in his case in a highly geometrical way and through an interpenetration of natural and artificial elements, in particular of natural landscape and urban elements. So, while the others put artefacts within a natural space, he puts nature within an artificial space.

---

[8] Cf. Marinetti's work of 1912 *Le Monoplan du Pape* (translated in 1914 in the Italian *L'aeroplano del Papa*) and Ardengo Soffici's poem of 1915 *Aeroplane* (*Aeroplano*), which are readable now in Sanguineti (1971: 550–566 and 646–650) respectively (but the first only in part). On the development of Aeropainting see Marinetti (1983: 628–636); cf. Crispolti and Sborgi (1998: 22–23, 216).

6

The question may be posed why an innovative representation is an interesting option. I think it is such for it requires coming back to our previous views and revising them, which is a more difficult task than applying old views to new things. But then a further question may be posed concerning the representation of nature in particular. The question is whether such an artefactual representation *of nature* is an interesting representation of nature. I will offer some reasons to such a conclusion and wonder whether those reasons make it even more interesting than the traditional representation of nature in natural terms. One may claim that a natural representation of nature is better and permits a better appreciation of nature itself. I will resist such a claim and remark that an artefactual representation of nature is quite (or even more) interesting. Why? For two kinds of reasons. In the first place, for aesthetic reasons. In the second, for ethical reasons.

7

Starting from the aesthetic reasons. It has been recently claimed that the expressive and representing powers of poetry are exerted using naturalistic terms. "What we usually call poetic imagery tends to be derived from myriad fundamental images of nature, such as the rising moon or setting sun, rushing rivers, wind-blown branches of trees, and all the distinctive movements and appearances of animals. We are nature's fabulists, and it seems that no poet can write strongly without recourse to natural imagery."[9]

But, restating the previous argument, I think that the artefactual representation of nature is quite (or even more) stimulating and suggesting. For it challenges our established ways of looking at things. It requires a stronger effort in imagination and combination of elements. This is confirmed, as to poetry at least, by the fact that, in the history of literature, poets have made recourse more frequently to the conservative way of representing than to the innovative, and, concerning nature in particular, more frequently to the natural than to the artefactual representation of nature. So, it is a way deserving of further exploration and experiment.

Furthermore, were the innovative representation abandoned or neglected, the ways of looking at nature and natural spaces in our technological era would risk to confine to some sort of Arcadian fancying or

[9] Fletcher (2004: 4).

even dreaming of the return to golden age: looking at old things with an old eye, something which is hardly exciting.

To sum up. These are the aesthetic reasons in favour of the artefactual representation of nature: (a) it is quite stimulating and challenges our established views (which is a reason on the principle that, in art, doing what is stimulating and challenging is good); (b) artists have made recourse to it less frequently than to natural representation (which is a reason on the principle that, in art, doing what many have done is bad); (c) the artist must say something of his time, of the problems and issues of the world he lives in.

8

Moving to the ethical reasons. Since a representation of nature in artificial terms helps in seeing what is at stake in manipulating nature and the environment, the artefactual representation may be preferred to the natural representation of nature. The focus of the former is on old things seen with a new eye, while the focus of the latter is on old things seen with an old eye. If that is true, artefactual representations, by virtue of their contrastive combination of nature and artefact, permit to focus on what is eventually lost in manipulating nature (what can hardly be done with a natural representation of nature). Let me add that this should not be intended as a general condemnation of technology: man has always manipulated nature and what he has gained is *in general* preferable to what he has lost (not in all cases of course). However, artefactual representations can be intended as warnings against irresponsible and dangerous manipulations.

We can provide a further argument in favour of artefactual representations, inspired by John McDowell's philosophy. The argument takes the artefactual representation to be a grasping of our second nature, or our cultural nature, of ourselves not only as sentient creatures but also as agents. In McDowell's picture, the notions of first and second nature serve to grasp the complex character of human understanding and rationality, and the relations between what he calls the natural realm of law and the space of reasons. The notion of second nature, in particular, integrates a form of naturalism which does not reduce intellectual spontaneity to perceptual receptivity, nor conceptual capacities to natural powers. What McDowell argues is a naturalism of second nature, an account of our nature as constituted by both inborn potentialities and upbringing.

"Our nature is largely second nature, and our second nature is the way it is not just because of the potentialities we were born with, but also because of our upbringing, our *Bildung*. Given the notion of second nature, we can say that the way our lives are shaped by reason is natural, even while we deny that the structure of the space of reasons can be integrated into the layout of the realm of law."[10]

Kant's lack of a pregnant notion of second nature, claims McDowell, explains his difficulties in providing a right conception of experience and his difficulties in accounting for the real connection between concepts and intuitions.[11] As to our purposes, artefactual representations help in seeing ourselves as agents and producers of artefacts. They help in seeing our second nature as embedded, but not reducible to, our first nature. This was the sense of Pascal's question I referred to in the very beginning: taking culture as second nature. And this can be extended to artefacts: taking artefacts as (products of our) second nature.

To sum up. These are the ethical reasons in favour of the artefactual representation: (d) it helps in seeing what is at stake in manipulating nature and the environment; (e) it is in tune with the grasping of ourselves as second nature.

9

For aesthetic and ethical reasons (a)–(e) I claim that there is something interesting—and even important—about innovative (artefactual) representation of nature. Looking at nature and spaces with the eyes of our time is more stimulating, suggesting and challenging than looking at artefacts and nature with the eyes of a previous time. First and second nature articulate our being and doing, which are neither in conflict nor separate. But looking at first nature from second nature, rather than the opposite, challenges and improves our understanding of the world and of ourselves, and, eventually, helps us to engage with nature.

References

Bonito Oliva, R. and Cantillo, G. (eds.) (2000) *Natura e cultura* Napoli: Guida.

Crispolti, E. and Sborgi, F. (eds.) (1998) *Futurismo. I grandi temi 1909–1944* Milano: Mazzotta.

Cendrars, B. (1967) *Du monde entier. Poésies complètes 1912–1924*. Paris: Gallimard.

[10] McDowell (1996: 87–88).

[11] *Ibid*. 97–98.

—. [trans. by R. Padgett] (1992) *Complete Poems* Berkeley: University of California Press.

Fletcher, A. (2004) *A New Theory for American Poetry. Democracy, the Environment, and the Future of Imagination* Cambridge: Harvard University Press.

Govoni, C. (2000) *Poesie 1903-1958* Milano: Mondadori.

Marinetti, F.T. (1983) *Teoria e invenzione futurista* Milano: Mondadori.

—. (2002) *Selected Poems and Related Prose* New Haven: Yale University Press.

McDowell, J. (1996) *Mind and World* Cambridge: Harvard University Press.

Mengaldo, P.V. (1978) *Poeti italiani del Novecento* Milano: Mondadori.

Millikan, R.G. (2004) *Varieties of Meaning* Cambridge: The MIT Press.

Papineau, D. (2003) *The Roots of Reason* Oxford: Oxford University Press.

Sanguineti, E. (1971) *Poesia italiana del Novecento, vol. II* Torino: Einaudi.

CHAPTER SIXTEEN

INTERIOR STYLES / EXTRAVAGANT LIVES GENDERED NARRATIVES OF SENSI/ABLE SPACES

MARIA TAMBOUKOU

Introduction

In this paper I am looking into sensible and sensable perceptions of urban and rural spaces, focusing in particular on how gendered power relations intervene in their configuration, renegotiation and ultimately representation. My discussion revolves around the epistolary narratives and paintings[1] of Gwen John, a Welsh artist who studied at the Slade School of Art in London but mostly lived and worked in Paris, in the first half of the twentieth century.

What I have found challenging and intriguing in Gwen John is the particularly spatial way that John has been perceived and represented, emerging as either a figure *par excellence* of the interior[2] or as an exemplary paradigm of the impossibility of the *flâneuse*.[3] Despite their differences, these representations work towards pinning down female subjectivities within statically conceived spaces and places, be they the room or the studio, the boulevard or café, the city or the country. What I will suggest in this paper is that John's spatiality should be charted within a Deleuzo-Guattarian image of thought, a dynamic process opening up planes of smooth spaces that "privilege difference rather than identity,

[1] Gwen John's paintings can be viewed on different websites, see the Bridgeman Art Library for a comprehensive collection http://www.bridgeman.co.uk. The references to paintings are followed by catalogue numbers from Cecily Langdale, *Gwen John: With a Catalogue Raisonné of the Paintings and a Selection of the Drawings.*

[2] Langdale and Jenkins (1985).

[3] Wolff (1994).

movement rather than stasis and change rather than what remains the same".[4]

In attempting to chart lines of flight of John's spatial constitution as a female subject of modernity, I have been quite intrigued by the force and passion that she both paints and writes about interior and exterior places, possibly because I have always been drawn to narratives of space, of how we tell and write stories about the meaning and significance of space and of those related concepts that comprise what Soja calls "the inherent spatiality of human life".[5] According to Michel de Certeau, "stories carry out a labour that constantly transforms places into spaces or spaces into places";[6] moreover, places are constituted in the very writing of 'space narratives', which further 'traverse' and 'organize' them.[7] Narratives are therefore instrumental in discussions revolving around the spatial constitution of the female subject.

A narrative approach to the analysis of spatial practices however is inherently limited. As de Certeau has suggested, "stories provide the decorative container of a narrativity for everyday practices [although] they describe only fragments of these practices; they are no more than its metaphors".[8] Edward Soja has further pointed to "the incapacity of language, texts, discourses, geographies and historiographies to capture fully the meanings of human spatiality",[9] while Nigel Thrift's non-representational approach foregrounds practices and performances rather than discursive representations in making sense of human geographies.[10] In highlighting the limitations of textual inscriptions of space and spatial practices and relations, Doreen Massey has most forcefully suggested that space—very much like time—is impossible to be pinned down, stabilized and textually represented.[11]

Given these limitations, of the textual inscriptions of space and place, what I will experiment with in this paper is a genealogical schizoanalysis of John's narratives drawn on her letters as well as her paintings. In following genealogical analytics,[12] I am problematising the universalism and quasi-inclusiveness of the very notion of the category of women and

---

[4] Lorraine (2005: 159).
[5] Soja (1996: 1).
[6] De Certeau (1988: 118).
[7] Augé (1995: 84).
[8] De Certeau (1988: 70).
[9] Soja (1996: 57).
[10] Thrift (1996).
[11] Massey (2005).
[12] Foucault (1986).

in avoiding normalizing inclusions, I am focusing on textual and visual narratives of Gwen John, a 'grey' figure in the history of art that has nevertheless attracted much attention in the feminist project of retrieving women artists from the obscurity in which they had lain for so long. As has been convincingly argued, 'hidden in history' women artists further remained in the margins of the bohemian circles of modernity.[13] What I will argue however in this paper, is that John is further marginalised within the margins, constituting herself—mainly through her letters—as a troubling figure, difficult to fit within the traditional boxes of the social and spatial formations of modernity. The way her persona has been portrayed is indeed an effect of the difficulties and troubles John—as an unruly female figure—has caused in structuralist theorizations of gendered spatial relations.

The genealogical strategies of this paper will further make connections with the Deleuzo-Guattarian proposition of schizoanalysis, an approach that challenges the integrity and indeed possibility of any centrally organized assemblage, be it the organism, the state, the unconscious or in the case of this paper, the sequential order and coherence of narrative texts.[14] What I will experiment with instead is looking into snapshots of spatial practices, fragments rather than coherent and closed space narratives. After all openness and fragmentation have been theorized as repetitive patterns in women's autobiographical narratives offering views 'from elsewhere' as Gillian Rose[15] has argued, into the panoramic patterns of masculinist geographies. These fragmented and open space narratives instead of pinning down the subject into striated and rigidly divided spaces form a series of planes wherein the spatial constitution of the female self is being considered always in making connections with other actual and virtual possibilities of becoming a subject. While these virtualities remain beyond representation in discourse, they do however constitute radical political imaginaries of future becomings. What I will therefore propose is that a genealogical schizoanalysis of John's narratives drawn on her correspondence as well as her paintings chart a different cartography of her spatiality wherein the fluidity of space and the fluidity of the female subject make nomadic connections. As Lorraine has succinctly noted, the nomadic subject of Deleuze and Guattari's philosophies, leaves behind conventional notions of space and time, orienting herself in heterogeneous

---

[13] See amongst others, Parkhurst (1994); Parsons (2000); Perry (1995, 1999); Pollock (1988); Ryan (1994); Wilson (1991, 1995); Wolff (1990, 1994).

[14] For a discussion of schizoanalytics see amongst others, Colebrook (2002: 132–138).

[15] Rose (1993).

blocks of space-time, not necessarily linked into rational wholes or underlying causalities.[16]

Charting John's spatiality

In making a cartography of Gwen John's spatial constitution as a modern woman and artist at the dawn of the 20th century, I have followed Doreen Massey's[17] suggestion for a non-static route of perceiving and analyzing spaces and places. Indeed a wide range of spatial themes, concepts and problems have emerged from this genealogical schizoanalysis of John's narratives. Not surprisingly, given the spatial tactics of the artistic milieus of her era, John's cartography is marked by displacement: indeed she chose to dislocate herself twice: first by leaving Wales to go to London to study at the famous Slade in the shade of her already notorious brother Augustus and then by moving to Paris, never returning home.

I have been particularly interested in tracing John's lines of flight from London to Paris, while narrative schizoanalysis as a research tool has enabled me to trace non-linear paths and routes in John's spatial constitution as a woman and as an artist. Indeed her escape to Paris emerges as an interesting episode in the sensible/sensable conceptual pair. I have seen this displacement as an effect of the unbearable heaviness of the bohemian circles of London within which her brother Augustus had already become notorious. The way London was sensed by John was non-sensical, unintelligible, almost incomprehensible amongst the bohemians. According to her biographer, in the four years preceding her move to Paris (1888-1903), John was living in a London basement where "no ray of light ever entered".[18] Her obsession with literally living underground in the company of her cat puzzled her friends who further marginalized her by labelling her as a recluse. Thus, as a young woman trying to pursue her artistic aspirations and live independently, John had to leave behind the familiar spaces and places of London. As she was writing in her notebook on July 7th, 1923:

> You are free only when you have left all.
> Leave everybody and let them leave you.
> Then only will you be without fear.[19]

---

[16] Lorraine (2005: 159).
[17] Massey (2005).
[18] Chitty (1981: 49).
[19] Lloyd (2004: 130).

By escaping the tyranny of the freedom of the bohemia, John undertook some wild walking adventures in the French countryside which ultimately took her to Paris. John has indeed written long and lively letters to her life-long friend and fellow student at the Slade, Ursula Tyrwhitt about her walks in France in the company of her friend Dorelia McNeil in 1903. As her biographer recounts, literally rephrasing John's correspondence during this period:

> They earned their living … in the cafés. They would arrive, dusty and laden, at the principal bar of a village and offer to draw all comers for two francs a head (the price of a dinner). If the client did not wish to be drawn they would sing to him (for this purpose they practiced singing as they walked) … When they left that café late at night they were pursued by drunks to the haystack under which they hoped to sleep … The nights were cold under the stars … They lay on each other to keep warm, with their portfolios on top of them. There were sometimes strange noises in the dark … They always woke to a row of men staring down at them …[20]

Surely, covering one hundred and fifty miles on foot within a month, sleeping rough and drawing and singing in cafés for a meal does not exactly resonate with John's image as a recluse, particularly when after having spent three months of 'ordinary' life in Toulouse in late 1903, her letters were reflecting her nostalgia for a nomadic style of life:

> We have been here so long now; the time of trudging from village to village seems a long way off. It is very pleasant here, the people gave us plenty of work but I have not done much besides my picture; We shall never get to Rome I'm afraid—it seems further away than it did in England … Our adventures are not now of such a thrilling nature as we live in a town and have a room and sleep in like any bourgeoisie.[21]

Indeed, throughout her correspondence to her friend Ursula Tyrwhitt and to her lover and mentor Augustus Rodin, John is trying to make sense of herself as a woman and as an artist, constantly moving in a complex network of relational emplacements, wherein the interior, the urban and indeed the rural spaces of modernity blend and blur in the spatial constitution of the female self. In thus charting the cartography of John's nomadism I have created three planes of smooth spaces: a) the room and the studio, b) the street, the café and the public garden and c) the countryside, the river, and the sea. I have used these planes as heuristic

[20] Chitty (1981: 57).
[21] (NLW MS 21468D, ff. 7–8).

analytic devices, a kind of conceptual spatial configurations to enable the discussion of this paper.

Plane one: the room and the studio

> I have not been posing to-day, so I washed the floor and everything is brilliant in my room, do come and see my room soon—the floor is of red bricks, and there are tiles in the fireplace and two little grates and all the furniture is brand new, light yellow … I've got an armoire a glace which is a wardrobe with shelves and a glace front, and white lace curtains at the window—yes I am Parisienne and I feel more at home with the French than the English now, I think, at least the English over here.[22]

As this colourful topology of her room in a letter of 1906 to Charles McEvoy illustrates, John wrote extensively about it in her letters. Her room further became the theme of a series of interiors.[23] However, this 'passion for her room' is much more complicated than just 'a desire for an interior life', a line which having been taken out of the context of one of John's letter to her friend Tyrwhitt has become central in a range of critical commentaries about her life and work.[24] As I will further argue this 'desire for a more interior life' is not necessarily linked to interior spaces; it is related to solitude, a line of flight in the cartography of John's spatiality.

Being a rich theme of both her epistolary and visual narratives, John's room is nevertheless interwoven within a complex structure of what Foucault[25] calls 'relations of emplacement' and is indeed constantly being transformed into a variety of spatial levels, wherein inside and outside boundaries blend together in the spatial constitution of herself. However, this complex interweaving of relations of emplacement is still structured by long-held established oppositions and binaries between the private and the public, the inner and the outer, the social and the personal.[26] It is I

---

[22] (NLW MS 23874C, ff. 89–90).

[23] See, Langdale (1987: 15, cat. n. 17, 18, 24, 25).

[24] In this letter, written John actually writes to Tyrwhitt about her decision to live a solitary life : 'You have given me so much encouragement, Ursula you seem to agree with my decision to live as much as I can in a way that I have collectedness of thoughts … As to whether I have anything worth expressing that is apart from the question. I may never have anything to express, except this desire for a more interior life' (NLW MS 21468D, ff.72b–73).

[25] Foucault (1998: 176).

[26] Foucault has pointed out to the enduring domination of these binarisms (1998: 177), but it is mainly through a rich body of feminist research that the oppositions

suggest as an effect of these spatial divisions that the discourse of John's 'desire for an interior life' has emerged.[27]

In painting herself within her room, John has further offered different and sometimes contradictory portraits. In two paintings, A Lady Reading (c.1910–1911) and Girl reading at the window (1911) she presents herself as a spiritual figure, the image of the intellectual woman reading her book by the open window.[28] As its title indicates, in 'The Artist in her room in Paris' (c.1907–1909) she assertively presents herself as the artist sitting in her room, perhaps having just returned or maybe ready to go out as her hat is prominently displayed on her bed.[29] Indeed a hat and a parasol are often depicted in her interiors probably as signs of a woman who keeps moving in and out as part of her daily routine, as in 'La Chambre sur la Cour' (c.1907–1908). In this painting John is offering a domestic scene of a woman simply sitting and sewing.[30] While this is the figure of the domestic woman par excellence in the company of her cat, the banded hat, hanging on the wicker chair, reminds the viewer of the woman's life outside domesticity. As art critique, Alicia Foster has pointed out, John's interiors "contain signs of life beyond the walls of the rooms"[31] and it is from her paintings and writings that narratives shattering the "cultural image of the artist as a recluse" have also emerged.[32]

John's self as a woman and as an artist thus emerges as a figure of the threshold, moving in and out of smooth spaces in a cyclical repetition of differentiations, a becoming between multiplicities: the intellectual woman, the spiritual figure, the domestic woman, the art student that still experiments with her art, the assertive artist, the new woman. John is what she becomes, mapping the contradictions and tensions of pure difference in writing and painting herself in space.

In this process of spacing herself, her room would further become a professional place, where John would meet artists looking for a model and more rarely prospective buyers of her work. In this light, her small Parisian room would also become a space for social encounters and

---

and paradoxes weaving around the private/public dichotomy have been discussed and analysed. For an illuminating review of this research and problematics see amongst others. See Scott and Keates (2004), amongs others.

[27] See in particular, Langdale and Jenkin's publication for the 1985 Barbican Art Gallery exhibition, entitled: *Gwen John, An Interior Life*.

[28] Langdale (1987: 38, cat. n. 25).

[29] *Ibid*. 30, cat. n.18.

[30] *Ibid*. 34, cat. n.15.

[31] Foster (1999: 49).

[32] *Ibid*. 6.

networking practices with women artists from all over the world, flocking into Paris at the time to study and work. In the following letter dated between 1906 and 1907, John writes about a strange socialist woman who had visited her to arrange modelling sessions for a Russian artist:

> … a woman that I had never seen before entered. She was big with grey hair dressed somehow like a socialist I believe … the woman … began talking about troubles in Russia and about her life, she was a political writer … She admired my room very much and I showed her the kitchen and the view from the window (which is very beautiful, particularly at night, my Master has not yet seen it in dark) … In the end she left … and she also said: 'I would be very pleased to see you, I would also like to see your soul'! Oh Julie, I don't know what to say o this, neither do I know what she wanted, but it's a way of speaking I believe between socialists, so I promised to go and see her one evening…[33]

The above extract resonates the effects of dramatic political events and changes of John's times: the visit of a Russian socialist political writer fills her room with new excitements and ideas about how other women live, think or even dress. John feels drawn by the uncanniness of this strange encounter and writes a long letter about it. Her room and the world open up to each other and make connections. Despite the odd professional encounters as the one recounted above however, John never really perceived her room as a workplace while living in Paris, although in her letters to Tyrwhitt she had expressed her desire for a studio, a necessary place for her identity as a professional artist to be established. Thus on May 6th, 1910, she was writing:

> What I should like to do is to earn enough to take a studio, when you come over, with you and work there. And that must happen within a year at least.[34]

Indeed, it was only in 1910 after John had moved to Meudon, a suburb near Paris where Rodin also lived, that she kept her room in Paris as mainly a studio which she would visit daily to work. It is quite interesting that establishing her Parisian room as a studio occurred at the same time that she was introduced to the American art collector John Quinn who became her patron. The small apartment she took in Meudon would ultimately become both her home and her studio when she left the Parisian

---

[33] (MR\MGJ, B.J4\Letters to Julie 1906–1907). All translations of John's letters in the Rodin Museum Archives are mine.
[34] (NLW MS 21468D, f.44).

room in 1918. By that time many things had changed in her life, Rodin had died and it was mainly from this period onwards that John devoted herself to her work, her career boosted and she firmly establishing herself as an artist, exhibiting both in Paris and in London.

Her room is therefore often represented both in her letters and paintings as her corner in the world, a place where she could hide and disengage herself from the turbulence of life; a space to dream, to read, to contemplate, to write letters, to cry, despair and be happy. It was further an experimental space for her art to be developed and sometimes a social space to receive her friends and negotiate work both as a model and as an artist. The thread however, that connects the variety of spatial levels that her room was constantly generating while living in Paris, was her erotic passion for Rodin:

> My room is very neat I like it much in the evening when there are shadows in my room … I like days like these when I don't talk to other people. I am happier when I can think of you in tranquillity and read my books. I was intending to design it in the light of the lamp this evening, but I like it more talking to you know …[35]

In this extract from an undated letter to Rodin, John's depiction of her room is filled with the shadows of the remains of a very busy day 'outside' and it is indeed very different from the colourful picture of her letter to McEvoy cited above. John used to write to Rodin in the evenings as a way of reflecting upon her thoughts, feelings and deeds at the end of the day—a Stoic technique of meditation *par excellence*. The tranquillity of her room in inciting internal thoughts is depicted both in her paintings[36] and in her letters. John constantly wrote about her room in many of her love letters to Rodin. Interestingly however, feminist discussions of her passionate attachment to her room have omitted the erotic forces that indeed saturate John's written and visual narratives around her room. In writing about John's joyful attachment to her room, Mary Taubman has drawn on an extract from one of her letters to Rodin: "My room is so delicious after a whole day outside, it seems to me that I am not myself except in my room"[37] while Janet Wolff has used Taubman's citation in her discussion of John's love for her room in contrast to her "active hatred" of "being out on the streets of Paris".[38]

---

[35] (MR, MGJ, B.J3, undated letters without name or address).
[36] Langdale (1987: 30, cat. n.17).
[37] Taubman (1985: 18).
[38] Wolff (1994: 116).

I was particularly interested in the above extract, having perceived it as an explicit spatial articulation of John's self-realization, which–as it seems–is directly related to an interior space: her room. However, working myself in the archives of the Rodin Museum, where John's correspondence to Rodin is housed, I was surprised to read the rest of this sentence and puzzled by its omission from Taubman's and consequently Wolff's quote. In the omitted extract John was linking existential consciousness both with her room and with Rodin, actually writing that

> It seems to me that I am not myself except in my room and in my master's studio' [my emphasis], further adding that 'My Master is the centre of my Paradise'.[39]

The whole letter in fact, is written at the end of a long tiring day that seems to have been 'wasted' for John as she had to pose for other women artists:

> I was posing for the whole day! For Mrs Smidt in the morning and Miss O'Donnel in the afternoon. I had lunch outside, in a restaurant and it seems to me that I didn't have time to be anything else than a machine for the whole day.[40]

Working as a model was indeed John's way to make ends meet so that she could support herself while working as an artist and it was actually as a model that she had met Rodin in the first place, posing for a monument to Whistler.[41] Her letters often depict her disillusion and frustration about having to work as a model, with the exception of course of her sessions with Rodin. Coming back to love her room after a whole day out working, does not therefore reflect any specific hatred of being out on the streets of Paris as Wolff maintains.[42] On the contrary as I will show later, John loved being out and according to her letters in walking she felt happy and rejuvenated. Moreover her love for her room runs parallel in her discourse to her love for her master's[43] studio.

---

[39] (MR, MGJ, Box J4, Letters to Julie, undated, non-localisable).

[40] *Ibid.*

[41] In 1904, Rodin posed John as a Muse for a monument to Whistler, commissioned by the International Society of Sculptors, Painters and Gravers. The monument was never cast in Bronze, since it was turned down.

[42] Wolff (1994: 116).

[43] Rodin was often addressed as 'the master' by his circle. Rilke for example who worked with Rodin as his secretary would always address him as 'My dear Master', a salutation that John would also use to open her letters to Rodin. See the

In writing to Rodin, her room is therefore reconfigured as an erotic space, constituting an emplacement related both to the domestic realm of femininity—as depicted in the letter to McEvoy above—but also to the creative sphere of Wolff's 'room of one's own'—as mostly expressed in her letters to Tyrwhitt. Her room was indeed the place sheltering brief erotic encounters and beyond these, John's pleasures, pains, suffering and dreaming of her agonising and passionate love. I think it is not accidental that after Rodin's death in 1917, John never painted herself in her room again, although she did interiors of her top floor studio in Meudon. In any case painting her room and writing about it in her letters went far beyond merely representing 'the feminine space of the home' as Wolff has commented.[44] Seen in the context of her extended correspondence to Rodin and not just captioned by some quotes taken out of her letters, John's interiors are filled with erotic tension and anticipation, inciting contemplation and gearing forces of deterritorialization. Eros in Deleuze's philosophy, holds the secret "of an insistence in all our existence";[45] in this light, it is through the triggering of the erotic that John's lines of flight are set into motion despite the reterritorializations that will inevitably occur within the patriarchal and heterosexual segmentarities of her (and indeed our) times.

In John's letters however, a room to love would often be transformed into a room to hate, and forces of deterritorialization would open up the spaces of the outside inciting a desire for an 'exterior life' deployed in the streets, cafés and public gardens of Paris.

Plane two: the street/ the café/ the public garden

Clearly John has done much more than simply writing and painting within her room. Despite the gendered restrictions imposed upon her mobility she has indeed been out and about in the urban spaces of modernity and she has both painted them and written about them: "My Master, I return from a walk I took in the dark streets, which did me good, since I was so sad".[46] John was quite aware of the risks of such expeditions but she–like most of her contemporaries–would deploy specific spatial tactics in handling

---

recent publication of Rilke's letters to Rodin, which is actually entitled as 'Cher Maitre' in Rilke and Sabbatier (2002).

[44] Wolff (1994: 119).

[45] Deleuze (2004:107).

[46] (MR, MGJ, B.J3, undated letters without name or address).

gendered restrictions of her mobility: "I am obliged to walk very quickly, so that men will not talk to me, but in walking I am happy …".[47]

Given the financial difficulties that John—like many of her contemporaries—would face as a single woman artist in Paris, struggling to make ends meet, cafés or restaurants would also be transformed into places for reading and letter-writing, as the following letter of February 4th, 1910, written on a printed letterhead indicates:

> Taverne de la Brasserie Dumesnil Frères,
> 73 Boulevard, Montparnasse, Paris.
>
> Dearest Ursula,
> Isn't this a dreadful piece of paper? It is so cold in my room and I haven't the energy to light the fire so I've come here to write to you, there is a band here and a lot of startling ladies amongst the men but I have books and writing paper all over the table so they think me only a mad Anglaise.[48]

Clearly these public spaces would not always be perceived as friendly and welcoming for a single woman. John has frequently written about her frustration of being in public places as in the following undated letter, where she recounts her experiences of painting tram horses in a Parisian café:

> I have just returned from a café where I was designing the horses. A man dressed as a dandy came near and started talking to me. I had gone there to design and not to talk to him, so I told him that I could not understand French. To my surprise instead of going away on hearing that, he sat next to me, ordered a coffee and started talking to me […] I could not understand what he was saying because I was focusing on my design but I could follow some of his words. In the end, people would stop and look at us and I was feeling very agitated and could not concentrate. I could hear him saying that he was a journalist and that he was very annoyed with the conductors who were staring and laughing at us. I was designing the ears of a horse and he was saying that when I would design the conductors I should make them with donkey ears … I could not help laughing … but still I was very annoyed and I left since so many people were staring ...[49]

Laughing away but always going away in the end, were some of the tactics that John would deploy in surviving the urban spaces of modernity, while

---

[47] (MR, MGJ, B.J3, undated letters without name or address).
[48] (NLW MS 21468D, ff.38–40).
[49] (MR\MGJ, B.J4\letters to Julie, undated, non-localizable).

the discourse of being a stranger 'I do not understand French' was often used but not always successfully, since it was women artists all over the world that 'the men in the street' were after. Indeed, on many occasions of finding herself in public places, writing would become an agonistic medium to defend herself, 'have collectedness of thought' as she would often put it.

John's life cannot therefore be placed in either compartment of the private/public, inside/outside model. Neither could be argued that despite her possibility to access the public sphere, it was actually her room that she really wanted to be in.[50] What I have found intriguing in tracing narrative lines of her life both in her letters and paintings, is the interchange of interior and exterior spatial experiences in a kind of a cyclical recurrence, a Nietzschean eternal return of her 'will to solitude'. In this context, the city gardens often emerge as a heterotopic spatial configuration in the matrix of John's relational emplacements. As she was writing to Tyrwhitt in July 1904:

> We have been out for a walk it is quite late, the sky is a deep blue with some great clouds, the Luxembourg gardens looked so beautiful with no soul there so quiet and peaceful & the trees are so beautiful down the streets casually lit up as a lamp. I sometimes sleep in the gardens in a little corpse of trees.[51]

The Luxembourg gardens, a rather short walk from all the places, John used to live in the area of Montparnasse were thus a frequent destination. There were different occasions on which John would spend time in the gardens. Sometimes it was because she could not stand the unbearable heaviness of waiting for Rodin. At these moments, the room she loved so much was being rendered into a suffocating place she had to get away from. As she was writing to Rodin: "I desire you so much my Master that I can't take pleasure in my room now",[52] and in a different letter: "…I am now going to the Luxembourg gardens my Master … In the gardens I always think of you in tranquillity and I keep the thoughts that annoy me at a distance".[53] The Luxembourg gardens would thus recreate the calm atmosphere of her room and paradoxically enough would be reconfigured as outer spaces of intimacy wherein she would live her love and passion for Rodin without the annoying anticipation of his visit.

---

[50] Wolff (1994: 118).
[51] (NLW MS 21468D, f.14).
[52] (MR/MGJ, Box J3/undated letters without name or place).
[53] (MR/MGJ, Box J3/undated letters without name or place).

While working in Paris at the archives of the Rodin Museum, I followed John's footsteps from the various addresses to the Parisian boulevards and the Luxembourg gardens. It was a strange experience as I was attempting to recreate existential paths within different space/time conjunctions. As Williams has noted "each present, each life is connected to all others but to greater and lesser degrees of contraction"[54] and in this light Deleuze has suggested that "one life may replay another at a different level".[55] Finding myself replaying John's paths a hundred years later, I was really struck and perplexed by the power of her strange familiarity with *fin-de-siècle* urban spaces. How could a public garden be transformed into an intimate space sheltering a young woman's sleep? Again, John's spatial refrains could not become intelligible within the long held gendered dichotomies of space.

The urban streets and gardens as spaces of the outside would thus reactivate internal spaces—the spaces of our dreams, our reveries and our passions that Bachelard[56] has written about—opening up heterotopias for the female self to redistribute and indeed redefine herself in space. In this light, what I suggest is that the Parisian streets, cafés and public gardens—spaces of the outside *par excellence*— as spaces of the outside should be mapped and analysed as heterotopias in the context of John's epistolary narratives.[57] John's intimate practices—writing, reading, contemplating and even sleeping—deployed as they are within exterior/public and therefore inimical places for women, open up heterotopic spaces within urban emplacements, functioning as Foucault[58] has delineated between illusion and reality as difference. In this light, John's intimate practices can no more be registered within the outer/inner division; they rather expose its illusionary boundaries, which literally melt into the air. Getting out would not however been restricted to the urban spaces; they would be often extended to the countryside surrounding Paris and not rarely even further.

Plane three: the countryside/the river/the sea

In search of her love for the countryside, a recurring theme of her personal writings since the time of living in London, John would often set out for excursions in the surrounding countryside, taking the boat down the Seine,

[54] Williams (2003: 97).
[55] *Ibid*. 93.
[56] Bachelard (1994).
[57] For a discussion of heterotopias, see Tamboukou (2000).
[58] Foucault (1998).

walking, painting, reading and again sleeping in the woods when tired. As she was writing to Rodin:

> I did not sleep well tonight either and after having tried to draw, I finished my housework, took my book and went out to the country […] now I feel better since I have been out for a long walk in the country. It is strange how walking for long relieves my heart! ...[59]

Being out in the country was not only an exciting experience but a risky one as well. John has written extensively about dangerous encounters during her walking adventures in the countryside surrounding Paris. These risks however do not seem to have discouraged her love and enthusiasm about being in the country. As when moving in the urban places, John would deploy specific tactics of surviving rural dangers as her following undated letter to Rodin recounts:

> The place I most like going is near a lake, there is much greenery there, big spaces and groups of very big trees. I did not go far away from the path, for the sake of security and so I remained at a spot where people could see me. (This is because there are many men who always go around, but if you are in the open, they don't dare come near you). As for me, I ate first and then I slept.[60]

Sometimes however, she would overstep her tactics and strategies when her desire would overcome her fears. On such occasions she would be prepared to defend herself by all means should danger arise. In a letter to Rodin, John writes in detail about her adventure of going in the country near his house and choosing a room in a deserted building that was giving her a good angle for painting his villa and studio. In realizing that a man was watching her however, she got infuriated and forcefully challenged him to go away, not hesitating to throw big stones at him that could have badly hurt him. The following extract from her letter vividly depicts these moment of self-defence:

> After a moment of terror, I got very angry– I was crying out: 'What are you doing there? Get out immediately! … And then I picked up a stone … and I threw it at him …I also threw a sardine tin that I found around … he was afraid because there were many stones around… When I picked up a big pointing stone, as big as a hat and I was threatening him with it … he was saying …'let the stone or I will (I don't know what). I told him that I was going to throw it at him if he didn't leave immediately … and I was

---

[59] (MR, MGJ, B.J3, undated letters without name or address).
[60] (MR/MGJ /BJ3/Undated letters without place or name).

> pretending I was throwing it at him… I am glad that I didn't throw it though, since it would have hurt him a lot and I didn't really want to do that, but only to make him leave. I have to tell you that he was not a worker, but I think 'a Monsieur' … When it all calmed down, I continued with my painting.[61]

According to her letters, John would often find herself in confrontation with men and would use a variety of strategies and tactics to get rid of them. As her biographer has noted some times her practices of resistance would verge to the point of criminal assault.[62] She would ultimately override all difficulties however, given her love and passion of being outside, surrounded by wilderness. Being near the sea in particular was according to her biographer one of the greatest passions of her life.

Indeed, since the days of her life in Wales, John used to search for secluded and wild spots where "she could bathe naked off the rocks and swim far out to sea, excited by the fear that she might not be able to get back".[63] John had expressed this great passion for the sea in her letters long before moving to Paris. As she was writing to her friend Michel Salaman in the spring of 1899, while staying with her brother Augustus and his wife Ida in Liverpool:

> I bathe in a natural bath, three miles away, the rocks are treacherous there, and the sea unfathomable. My bath is so deep I cannot dive to the bottom, and I can swim in it- but there is no delicious danger about it, so yesterday I sat on the edge of the rock to see what would happen- and a great wave came and rolled me over and over – which was humiliating and very painful and then it washed me out to sea and that was terrifying-but I washed up again.[64]

In the last years of the First World War and while residing in a deserted mansion in Brittany where she could work in peace, John would take daily walks to the nearby beach, often diving "into the crashing breakers of the Grève des Vallèes".[65] As her biographer has further noted:

> She was fascinated by the steady succession of those waves and claimed that after nine little ones there always came a big one. In a note found

[61] (MR/MGJ/BJ4/ Fonds Judith Cladel).
[62] Chitty (1981: 105).
[63] *Ibid*. 30.
[64] Lloyd (2004: 21).
[65] Chitty (1981: 152).

> among her papers she declared that her heart also was a sea which had sad little waves, but every ninth one was great and happy.[66]

There is more order of course in the cyclical repetition of John's imagination than a Deleuzian image of thought would have synthesized, but still, the sea becomes a recurrent theme in the cartography of her spatiality. The sea after all, has been a quite fascinating spatial metaphor in a genealogical line of philosophical, artistic and literary imaginaries of the self. Lorraine[67] has discussed how in Irigaray's thought the sea expresses the material nature of reality in a register that goes beyond the discursive constraints of language. Taken as an extralinguistic element in Irigaray's thought, the sea of John's real and imagined experiences becomes a plane of consistency wherein her epistolary and visual narratives can be imagined floating like moving sands of her becoming other.

Gendered experiences of sensi/able spaces

In working with John's epistolary narratives and paintings, what I have suggested is that gendered experiences of lived spaces create conditions of possibility for her artistic sensibilities to be deployed along lines that keep transforming her and the spaces into which she constantly immerses herself. Both her letters and her paintings chart a unique map of her lived spaces within which she keeps redistributing and consequently reinventing herself and her relation to the world. In portraying women and indeed herself within *fin-de-siècle* spaces and places, John forcefully brings forward the strangeness of the familiar, alongside the uncanny intimacy of the distant, the strange, and the unknown. In this process of spatial and existential reconfiguration, boundaries between external and internal worlds seem to be blurred in ways that radically interrogate long standing dichotomies between her room and the street, the city and the country, the domestic and the professional, the public and the private, the inside and the outside. Moving in between relational emplacements of modernity, John follows solitude as a line of flight in becoming a subject, constantly reaching beyond the discursive constraints of the social outer and the psychological inner. In this light, her letters and paintings constitute narrative articulations of the spatial constitution of the female self as it surpasses the inner/outer division and can offer insights into gendered configurations of sensi/able spaces.

---

[66] *Ibid*. 152.
[67] Lorraine (1999: 44).

Archival sources

National Library of Wales, Archives, Gwen John's papers (NLW MS)
Rodin Museum, Marie Gwendolen John's boxes (MR\MGJ)

Acknowledgements

This paper developed as part of the project 'In the fold between life and art: a genealogy of women artists' (2003-2006). I am grateful to AHRC for funding its initial stage (2003-2004) and the University of East London for funding the visit to the Rodin Museum in Paris (May-June 2005). I would like to thank Mme Virginie Delaforge, archivist of the Rodin Museum Archives for all her help and support during my research visit there and the Archives for giving me permission to quote from John's letters. Finally I am grateful to Ms Sara John for giving me permission to get photocopies of Gwen John's letters to Ursula Tyrwhitt. This correspondence has now been published in Lloyd-Morgan, 2004 but the experience of working with the manuscripts has indeed been unique.

References

Augé, M. (1995) *Non-places, Introduction to an Anthropology of Supermodernity* London: Verso.

Bachelard, G. (1994) *The poetics of space* Boston: Beacon Press.

Chitty, S. (1981) *Gwen John* New York: Franklin Watts.

Colebrook, C. (2002) *Understanding Deleuze* London: Routledge.

Deleuze, G. (2004) *Difference and Repetition* London: Continuum.

De Certeau, M. (1988) *The Practice of Everyday Life* Berkeley: University of California Press.

Foster, A. (1999) *Gwen John* London: Tate Gallery publishing.

Foucault, M. (1986) 'Nietzsche, Genealogy, History' In P. Rabinow (ed.) *The Foucault Reader* Harmondsworth: Peregrine. (76–100)

—. (1998) 'Different Spaces [1984]' In P. Rabinow (ed.) *Michel Foucault, Aesthetics, Method and Epistemology, the essential works of Michel Foucault, 1954-1984, vol. II* Harmondsworth: Penguin. (175–185)

Langdale C. and Jenkins D.F. (1985) *Gwen John: An Interior Life* New York: Rizzoli.

Lloyd-Morgan, C. (2004) *Gwen John: Letters and Notebooks* London: Tate Publishing in association with the National Library of Wales.

Lorraine, T. (1999) *Irigaray and Deleuze: Experiments in Visceral Philosophy* Ithaca: Cornell University Press.

Lorraine, T. (2005) 'Ahab and Becoming-Whale: The Nomadic Subject in Smooth Space' In I. Bunchan and G. Lambert (eds.) *Deleuze and Space* Edinburgh: Edinburgh University Press.

Massey, D. (2005) *For Space* London: Sage.

Parkhurst Ferguson, P. (1994) 'The flâneur on and off the streets of Paris' In K Tester (ed.) *The Flâneur* London: Routledge. (22–41)

Parsons, D.L. (2000) *Streetwalking the Metropolis: Women, the City and modernity* Oxford: Oxford University Press.

Perry, G. (1995) *Women Artists and the Parisian Avant-Garde* Manchester: Manchester University Press.

—. (1999) 'The Parisian avant-garde and 'feminine' art in the early twentieth century' In G. Perry (ed.) *Gender and Art* New Haven: Yale University Press in association with the Open University.

Pollock, G. (1988) *Vision and Difference: Femininity, Feminism and the Histories of Art* London: Routledge.

Rilke, R.-M. and Sabatier, K. (2002) *Cher Maître, Lettres a Auguste Rodin* Broch: Editions Alternatives.

Rose, G. (1993) *Feminism and Geography. The Limits of Geographical Knowledge* Cambridge: Polity Press.

Ryan, J. (1994) 'Women, Modernity and the City' In *Theory, Culture and Society* **11**: 35-63.

Scott, J.W. and Keates D. (2004) *Going Public: Feminism and the Shifting Boundaries of the Private Sphere* Urbana and Champaign: Illinois University Press.

Soja, E.W. (1996) *Thirdspace: Journeys to Los Angeles and Other Real-and Imagined Places* Oxford: Blackwell.

Tamboukou, M. (2000) 'Of Other Spaces: women's colleges at the turn of the century' In *Gender Place and Culture* **7**(3): 247–263.

Taubman, M. (1985) *Gwen John: The artist and her work* Ithaca: Cornell University Press.

Thrift, N. (1996) *Spatial Formations* London: Sage.

—. (2004) 'Intensities of feeling: towards a spatial politics of affect' In *Geografiska Annaler B*, **86B**: 57–58.

Williams, J. (2003) *Gilles Deleuze's Difference and Repetition: a critical introduction and guide* Edinburgh: Edinburgh University Press.

Wilson, E. (1991) *The Sphinx in the City: urban life, the control of disorder, and women* Berkeley: University of California Press.

—. (1995) 'The Invisible Flaneur' In S. Watson and K. Gibson (eds.) *Postmodern Cities and Spaces* Oxford: Blackwell, Oxford. (59–79)

Wolff, J. (1990) 'The Invisible Flâneuse: Women and the Literature of Modernity' In J. Woolf, *Feminine Sentences, Essays on Women and Culture* Cambridge: Polity Press. (34–47)

—. (1994) 'The artist and the flâneur: Rodin, Rilke and Gwen John in Paris' In K. Tester (ed.) *The Flâneur* London: Routledge. (111–137)

CHAPTER SEVENTEEN

THE USE OF GEOGRAPHICAL ILLUSTRATION IN REPRESENTING THE RELATIONSHIP BETWEEN PEOPLE AND THE ENVIRONMENT

KENICHI NONAKA AND NOZOMI YANAHARA

Introduction

This study aims to represent the relation between people and the environment in the form of illustration. This 'relation' is made up of people's actions, related environmental elements, the site where the actions take place, reasons why people and the environment are connected and how close the connection is, and people's 'feelings' in taking the actions. The paper discusses how observation of these 'feelings' may reveal sensory aspects which are significant in understanding the relation between the environment and people. Also examined is the effectiveness of illustration, especially through manga (Japanese comics), as a means of visible representation.

Nonaka has carried out analytical field studies on the relation between the environment and people from the standpoint of geography and ecological anthropology.[1] As a professional manga writer, Yanahara has published many works of manga, most of which are for girls, such as *Otogi-banashi wo Kataro* (Talking about fairy tales),[2] *Maruichi teki Hukei* (Stories of a tiny robot named Maruichi),[3] *Torikae Hukaden* (The story of princess Huka's transformation)[4] and *Ketai-kun to Issyo* (Living with a transformed cell-phone).[5] In collaboration, Nonaka wrote and edited stories whilst Yanahara drew pictures illustrating various traditional

---

[1] Park and Nonaka (2003), Nonaka (2004 and 2005).
[2] Hakusensha comics (1994–1997).
[3] Hakusensha comics (1998–2000).
[4] Hakusensha comics (2004–2005).
[5] Hakusensha comics (2005).

practices concerning the relation between people and living things under the title of *Wakuwaku Ikimono Chirigaku* (Exciting Biogeography) in serial form. As exemplified in this series, the paper discusses how an illustration effectively shows the spatiality, temporality and composition of various elements, which are important in the relation between the environment and people, enabling readers to understand them clearly at a glance. Also showing that illustration may represent various intangible relations in a visible way, we will argue that this method of representation can present an argument. Furthermore, we will consider how effective illustration can be an interface in observing human feelings, as it may explain what is actually happening during a unique individual interaction in the field, allowing readers to easily understand and to empathize.

Special focus on senses and feelings in 'the relation between the environment and people'

Needless to say, the clarification of the relation between the environment and people is one of the most important subjects in the field of geography. In dealing with the subject, it is important to ask why any given action is taken or why a given result occurs, and try to understand people's actions towards the environment, the values and feelings of people upon whom the environment has an impact, and sociological and cultural contexts in addition to analysis of the direct impact and consequences of the relation. Thus, the relationship between people and the environment, from the viewpoint of this study, is arranged in table 4.

Table 4: The features of the environment as viewed by living things

Concreteness	Visible elements and human activity
Movement	Bio-systems and their activities
Interrelationship	Changeable/rotational network
Comprehensiveness	Nature Community/society Many cultures present together

The environment is complex, consisting of many elements. Each element is dynamic and therefore connections with other elements are constantly changing. People perceive elements and take actions toward them and, thus, a relation is formed between them. When such a relation is being formed with people, people look at and perceive the object, and at the same time have feelings. These feelings may be evoked according to human values or necessity. Studies concerning 'the relation' have focused on the cognition and perception of the environment, or human activities as people acting upon the environment. There have been, however, few geographical studies dealing comprehensively with the above. The former kinds of studies concerning cognition and perception have been presented as the theory of landscape, whereas the latter have concentrated on the results rather than the activities themselves. As a result, geography has turned away from the physical representation of the relation itself. The activities have been considered the domain of anthropology instead. Anthropology has observed these activities and made quantitative analysis of their results, including the types, amount, and yields of the activities. However, actions upon the environment should suggest human feelings, such as why people do what they do and how they feel when they do so. Exactly how and why human feelings may be involved depends upon their values, so these feelings may be an indication of people's motives for their activities and actions upon the environment. Such questions can connect perception with activities, which then may return our attention to their true appearance.

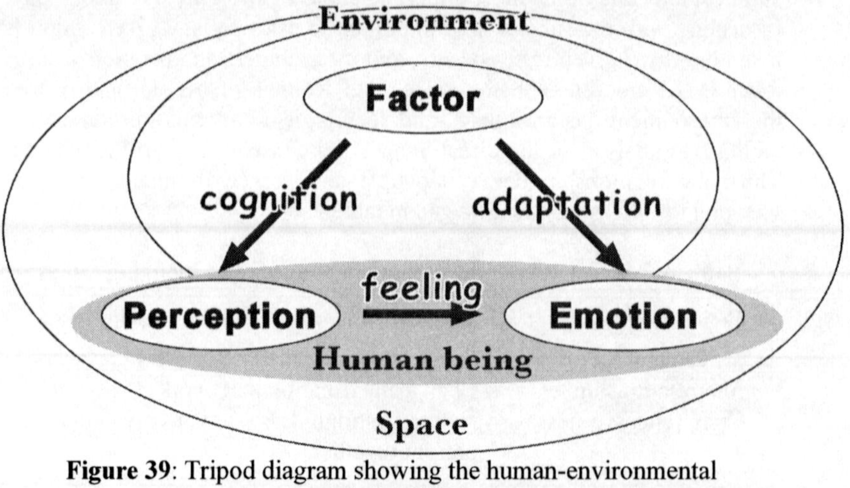

Figure 39: Tripod diagram showing the human-environmental relationship.

The above figure incorporates these human feelings and shows the relation between people and the environment. We perceive the environment by sensing its various elements. We take action following our decisions, which are based on senses and values. The relation between people and the environment is made up of physical things, events, and the interrelations amongst them which are structured in space and time and also combined with individuals' subjective feelings and emotions. Human feelings are very complex, for they are brought about by circumstances and history. In order to understand such complex relations, it is customary to read through long papers made up of linear expressions, but it is not easy to imagine the many environmental elements simultaneously, or picture in our minds the activities written in the paper.

The use of illustrations

In order for people to comprehend various elements interrelated in space, they should be represented by illustrations. Their advantages may be pointed out as follows: They represent intangible things visibly; they can represent actions; and they can simplify or emphasise things.

Meanwhile, when human-environmental relations are presented in the study, the first thing is to show the reality of the field based on physical things and events as the subject of the study. This study concerns the human feelings and emotions set out above as well as environmental elements, perception, and activities, all of which are necessary for forming the relationship. These elements are listed in table 5.

Table 5: Elements connected with human-environmental relations

Environment/information (Physical elements, people, things, sound, smell)
Feelings
Actions
Mental images (The environment perceived by people, mental reality)
Time (Passage of time, rhythm, history)
Circumstantial explanation (The locational aspect (to coin a phrase, 'locationality'): multi-layered-ness, multi-dimensionality and multi-directionality)
Changeable/variable scale (Changing viewpoint)

To compare the advantages of illustration with the elements concerning the relation of people to the environment, an illustration can show the elements of the environment comprehensively by giving an impression at a single glance and in its entirety. The same illustration also contains an individual relationship and individuality if the reader is attentive to the details. Therefore, the use of illustrations can give readers a general understanding of the whole picture before directing their careful attention to individual relations. Thus, illustrations may show the state of circumstances, and the changeability/viability of scale, at the time the subject perceives the environment or connects him-/herself with it. This may be difficult to show through conventional structure and network diagrams. In this regard, illustrations may prove to be more suitable.

At the same time, illustrations may also be useful as a means of describing the environment as interpreted in terms of feelings (sensible/sensable impression) in contrast to a means of showing perceptions as the results of conventional studies of landscape and pictures. By considering the feelings of people who form relations with the environment in terms of what is done (action), why it is done (awareness of action and interrelation), how it is done, and the location in which it takes place, it becomes possible to understand what significance people give to environmental elements. Further, if a representation focuses on feelings commonly shared among people, it can work effectively on the sentiment of readers. Empathy evoked among people would be an emotive power for moving people to take various actions. Representation in the form of illustration may prove that geography benefits modern society not only as analytical science but as pragmatic science, and even as art.

The use of manga as a representational technique in illustration

The techniques of Japanese manga are adopted as a means of illustration. This is because manga has accumulated various techniques for illustrating emotional expressions, movement of people and things, stories, and ways of evoking empathy in readers. Manga is said to trace its roots back to *emakimono* (picture scrolls) in the 12th century, and developed into *ukiyo-e* (woodblock prints) depicting everyday life in the Edo Period. Manga became more popular and widespread after appearing in newspapers and magazines from the Meiji Period[6] as a visual medium or a caricature/cartoon satirizing politics. With an increasing number of comic magazines published after World War II, manga has prospered as an industry. Today even in other countries, manga is recognized as a genre of Japanese art and culture, and the word 'manga' has become well known.

Manga is made up of pictures and a story that develops from frame to frame. A frame comprises pictures of characters, associated items, and the background illustrating the circumstances, along with written words depicting characters' speech and sentiments. What characterizes manga best is the many different kinds of lines which are used; lines with special effects and marks, as well as backgrounds drawn with abstract patterns.

[6] Shimizu (1997).

By using many different ways of drawing backgrounds and lines together with various marks and lines denoting multiple meanings, it becomes possible to draw clearly different expressions and represent the senses, impressions and psychology of characters.[7] Manga can not only describe the reality surrounding the hero/heroine but also depict his/her mental images and recollection of memories. Each page is made up of a series of frames and, by reading each in order, one is able to follow the story. This is the same as comics and cartoons, but manga utilizes more variations in the formation, size, shape, and layout of frames to express the psychology of characters and the pace of development of a story. Furthermore, using these techniques, manga can take readers back to the past. With the abstract function of arranging manga stories in time and space, the frames guide readers' eyes in an orderly manner through this time and space, just like the function of constructions in languages.[8] Table 6 shows the use of manga techniques.

Table 6: Features of cartoons and comics, or manga

Expression of complex human feelings
Representation of the atmosphere of a scene
Elaboration of peripheral things
Showing movements
Replacing one thing with something else
Guiding readers' eyes toward what we want them to look at

Figure 40: Expression of human feelings using manga techniques.

The above figure shows some examples. The basic model is a smiling face. A smile is a sign of accepting something that is happening. If a drop of sweat is added on the same face, Character I accepts it but at the same time is a little embarrassed. The face also looks dumbfounded. Character

---

[7] Natsume (1997).

[8] *Ibid.*

II is angry beneath the surface. Character III feels sad or shows unwillingness or reluctance. Character IV looks bewildered or fearful. All the faces show acceptance at least superficially, but added lines and marks express their complex internal feelings.

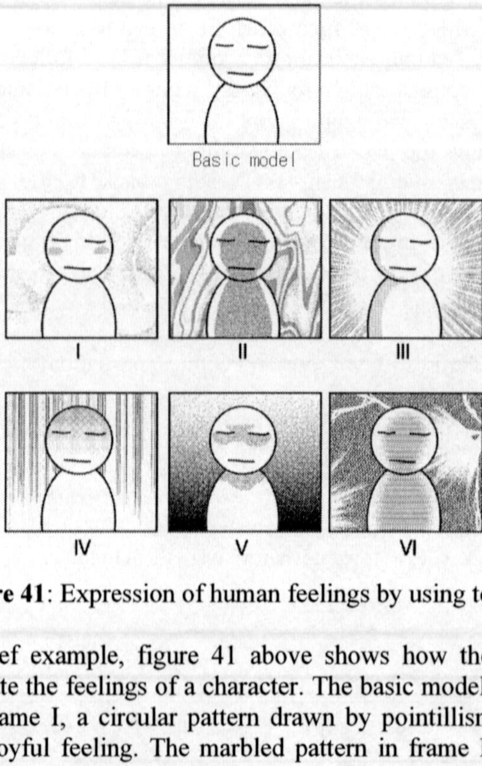

Figure 41: Expression of human feelings by using tones.

To give a brief example, figure 41 above shows how the background patterns indicate the feelings of a character. The basic model expresses no emotion. In frame I, a circular pattern drawn by pointillism expresses a buoyant and joyful feeling. The marbled pattern in frame II is a typical way of representing frustration. The pattern in frame III is often drawn to represent the moment a character discovers something. In frame IV, the pattern indicates low spirits, whilst the one in frame V suggests a grudge. The pattern shown in frame VI is used to express astonishment as well as anger. Combining background patterns and the expressions of characters enables more precise expression of more complex feelings and emotions. Moreover, these patterns sometimes suggest the atmosphere surrounding the character. Whether the pattern indicates the expression of a character or the atmosphere is understood intuitively by readers, as they follow the development of a story and framing. If the patterns suggest an ambience, each pattern means as follows:

the character is surrounded by warm and favourable atmosphere;
the character is surrounded by a confused atmosphere or confuses the scene;
the character attracts attention;
interest and attention in the character is lost;
the atmosphere is disquieting;
the atmosphere is tense. Manga has many other different rules concerning the representation of mental images.

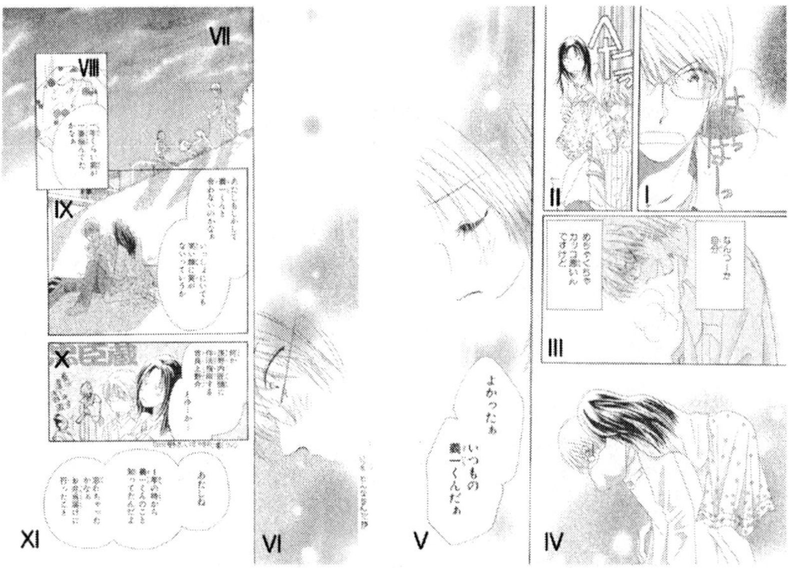

Figure 42: Example of a page from a manga story.
Source: Yanahara (2006)

Figure 42 above, shows some frames excerpted from an actual manga. As the Japanese language is formally written from top to bottom and from right to left, frames also proceed across the page from right to left. For convenience, the frames are numbered here. The book is bound between frames V and VI, and so the story moves onto the next page at this point. The backgrounds of frames I, IV, V and VI represent the feelings and emotions of the characters. The hero is uneasy in frame I. In frame IV, however, his feelings are reversed. He feels relieved and uplifted, for he was accepted by the heroine. Such a soft pattern as the one in the background of frame IV is often used when a character's heart is warmed. After reading frames I, II, III, and IV, which run from top to bottom, one's eyes are suddenly confronted by the rectangular frame V. Directing readers' eyes from the bottom to the top of the page implies a great change of the hero's emotion. Frame VII shows another scene. Dark shadows laid on the ground indicate that the scene shifts to an evening scene on an embankment. A quick shift from a close-up of a handkerchief belonging to the character in frame VIII to a wider angle in frame IX informs readers of the relative positions of the hero and heroine. Despite sharing the same background, frames VII and IX are divided into two independent frames. Frame VIII is inserted between frames VII and IX, providing readers with a chance to take a pause. This pause is experienced by readers as they visually follow the story from frame to frame.

Here, readers have a sense of time having passed for the hero and heroine. The pictures in the background of frame X are supplementary to the heroine's lines, where she talks about the scene of a historical drama. Clear depiction of the scene through pictures can convey her intended meaning even to readers who may not be aware of the scene. The important point we want to stress here is that, for the reader, these representation techniques follow commonly accepted rules. In other words, pictures drawn according to these rules can give readers a common understanding of the meanings of pictures. In the case of a work of art, it is the impressions of the artist that are represented, so the artist as a subject is reflected in the drawing. If the environment is represented in a work of art, it shows how the environment is seen by the artist.

In the study of human-environmental relations through fieldwork, however, it should be borne in mind that the subject is not an observer, but a person who is being observed and their actions. Objective description is necessary in presenting the geography of the environment and people. In this regard, manga is more suitable for objective description and explanation than artistic drawings and paintings that represent the sensibility of artists. Also required is an adequate means of representing a story, feelings evoked in people's minds, environmental information including elements or parts, and the interrelations amongst them. As they are equipped with a variety of representation techniques such as those explained above, the introduction of manga—especially those for girls, which have placed great importance on representing complex or multi-layered feelings—is effective in the spatial representation of the complex relations between the environment and people.

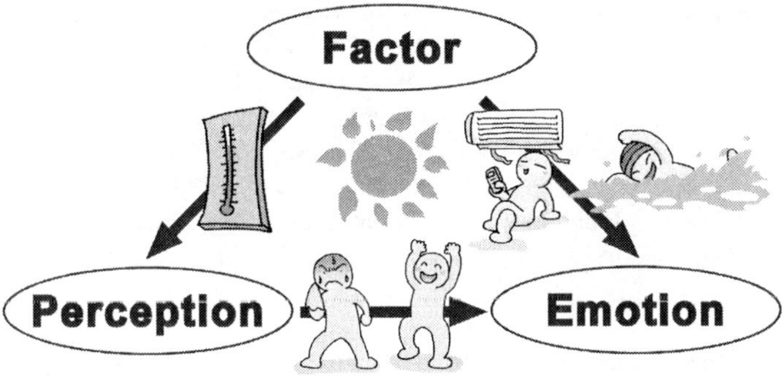

Figure 43: Human-environment relationship represented by illustration.

Figure 39, representing human-environmental relations above, has been adapted in the form of manga in figure 43 to specifically show the senses and actions of people. As to how people perceive and react to environmental temperatures, this figure can show the senses of hotness and different values concerning hotness, and also the resultant variations in action by the representation of emotions, in contrast to the former structural model in figure 39.

The representation of the relationship between living things and people: examples of traditional practices introduced in 'Exciting Biogeography'

There is a variety of customs and practices concerning the use of living things in the world. When people have direct relations with living things, they comprehend them with their sensitivity/sensibility while sensing first-hand their shapes, movement, temperature, touch, taste, palatability, and so on. Such sensitivity/sensibility causes people to have favourable or unfavourable feelings toward living things and create techniques for using them, through which they are related with each other. Then it leads to connections among various environmental elements which are spatially sustained.

The theme of the 'Exciting Biogeography' series was traditional practices and their relations with living things. How do people behave towards living things in diversity, change, and motion; how do they bring them into their life; how do they confront them? Such emotive feelings are called *Wakuwaku* in Japanese. *Wakuwaku* is a onomatopoeia, used to express human emotions concerning interest, discovery, surprise, excitement, and deep impression. *Wakuwaku Ikimono Chirigaku* (Exciting Biogeography) appeared 12 times from June 2005 through May 2006 in a monthly journal called *Geography* and was submitted by Nonaka (editing) and Yanahara (illustration). Actual cases were presented concerning the traditional use of natural resources, the breeding and cultivation of living things, the management of wild animals and plants, eating habits/food culture and the global environment. In the series, illustrations focused upon human feelings that are evoked in the process of establishing relations between people and living things (see table 7).

Table 7: The subjects and the objects taken up in 'Exciting Biogeography'.

Traditional use of natural resources
Yellow Jacket / raising, affection, community
Wild plant / gathering and distribution activities
Fish and shellfish / coastal fishery, continuation of cultural resources /
collecting aquatic creatures in the tideland,memory, sensory /
intuitive perception based on experience Small animals /
paddy field, cycle, biodiversity

Raising and culturing
Dogs / dogs allowed to run loose, comparison of cultures
Bulls / network for breeding fighting bulls
Pigs / indigenous domestic animals as cultural resources to be distributed in the market
Laver*(nori*) / cultivation industry dependent upon nature

Management of wild animals
Monkeys / damage caused by wild animals, harmonious coexistence and measures for dealing with the problem, and solutions

Eating habits/ food culture
Oysters / eaten raw or fresh, comparison of cultures

Global environment
Trees / flowers bloom, wither, and die, global warming

We thought that we should pursue answers as to how and where one can observe people's enthusiasm for living things. This is because, historically, this enthusiasm has been developed through human interactions with living things and the natural environment. Peculiar scenes were illustrated, consisting of the senses and feelings of characters and their associated environments, along with an analysis of practices. In the following section, three typical examples are used to explain a way of representing traditional practices based on sensory actions. Using these case studies, we show how to facilitate understanding among people and encourage empathy.

Case studies

The visualization of relationships and senses of values: the case of a family that raises yellow jackets. This is an example of how to represent abstract concepts, such as social systems, relations, and the sensory aspects of people, including feelings and gazes towards the object, and how to present a scene in order to suggest the characteristics of a place.

In the mountainous areas of Central Japan, a kind of wasp called a 'yellow jacket' (*Vespula flaviceps*) has long been savoured as a delicacy. Some people may react negatively to this eating habit, thinking that it is disgusting to eat insects. Others may get scared and wonder how it is possible to catch dangerous wasps safely. In these areas, however, the insects are considered very tasty and relished by the locals. Besides collecting wild wasps, people also bring small nests back home to keep in a box in the garden, where they carefully rear the insects. Moreover, some people even mate the next-generation queen wasp to breed them in their own house. Their enthusiasm provides the basis on which their relations with the wasp are sustained.[9]

For the uninitiated, such things may prove difficult to comprehend, although a little background information and further details should provide ample food for thought. Otherwise, people can hardly be expected to understand when they cannot even imagine what the wasps look like, what kind of food they are, or what they taste like. A full explanation, however, would take a great deal of time. If the reader has never heard of this kind of thing before, they may think it incredible that people actually eat insects and even raise dangerous yellow jackets, and may think that such talk should be avoided. The point is that these wasps are considered delicacies; local people raise them with the same care and attention as they do in bringing up their own children. Raising yellow jackets is an acknowledged practice in the local community, but there are also differences in the way each generation deals with yellow jackets. This is what we aim to share with the reader.

[9] Nonaka and Yanahara (2005).

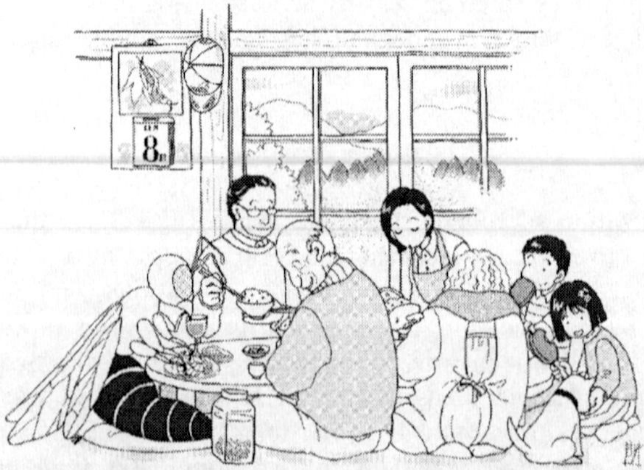

Figure 44: Dining with a yellow jacket queen.
Source: Nonaka and Yanahara (2005)

The above figure is intended to illustrate the environment of the area, the significance of the practice, and its sociological aspects. The picture attempts to explain the fact that yellow jackets are appreciated as food, and served on the family dinner table, a cultural practice in this area. Even though they raise yellow jackets with affection, their ultimate goal is to raise them until they are big enough to eat. It is hoped that the picture illustrates this point clearly. As this picture was drawn for our series, with Japanese readers in mind, there may be some points that are only understood within the Japanese context. It shows a typical scene of a Japanese family having a meal. Readers may notice that there are no wasps flying in the scene. A wasp as big as a family member, sitting at the table together with the family, is a symbolic representation of a queen yellow jacket kept in this village. The villagers say their feeling toward yellow jackets is one of affection, and that they care as much for them as they do for members of their own family, or lovers. This is why the dinner table was chosen as the setting for this picture.

To indicate where this practice takes place, the picture is drawn showing landscape characteristic of the area visible outside the window. Then, there are family members around the table. The family members are shown to accept the yellow jacket queen very casually, which is clear from their expressions and their behaviour. Such a yellow jacket, however, is a kind of food, which is suggested by the dishes on the table. Moreover, a difference can be seen in the interest shown in the yellow jacket by Grandpa, who loves the wasps and raises them, compared to the other family members' seeming indifference towards it. Grandpa raises them with such care as to give them much more sumptuous food than he gives to the family members, which is the way one would be expected to behave towards a lover. From a distance, Father observes the relationship between Grandpa and the yellow jacket, and Mother does not seem take much interest in it. The children accept such dinner guests as part of their everyday life, and eat dishes with the yellow jackets. In giving full consideration to the relationship between people and the environment, it is important to understand subtle differences in the interest shown in, and the treatment of something according to one's age and sex, even within a community. Items unique to the area are included in the picture in order to add a sense of reality to the representation of the life style and daily life of the local people.

People sensing signs and the environment: A case study of fishing in the tideland

The next example aims to show the relation and interaction with things that actually exist, albeit invisibly, and the traffic between the internal world and the reality of the external world. The subject of this study was women's gathering activities in the tide land on the island of Okinawa, located in the south of the Japanese archipelago.[10]

Fishing in the tideland is carried out on the night sea during ebb tides. Women go out to the tideland to collect fish and shellfish that are left behind in the low tide, using their bare hands or simple fishing tools. However the fish are not clearly visible, so the women find them by sensing their movement, smell, and touch. This touch is not direct, however, but is a product of the signs that they deduce from their past experiences and from the circumstances. These marine creatures are caught for food. Each fish caught is uniquely delicious. These fish and shellfish naturally have an economic value, but at the same time they are considered important as delicacies for the family, as well as gifts for relatives living far away. This story shows various memories evoked by the produce caught in the tideland and their taste, and that the women find and gather them by following signs. Local people retain memories from the historical process of developing the use of such resources. These memories and the circumstances of the tideland are combined to identify the existence of fish. This happens certainly in the world of senses. In an invisible world, their memories and the sensing of signs relate them to the

[10] Ikegichi and Yanahara (2006).

environment, which is made possible by the wisdom that is also developed, which means that the creatures are caught with wisdom. These facts are represented in the form of illustration in figure 45.

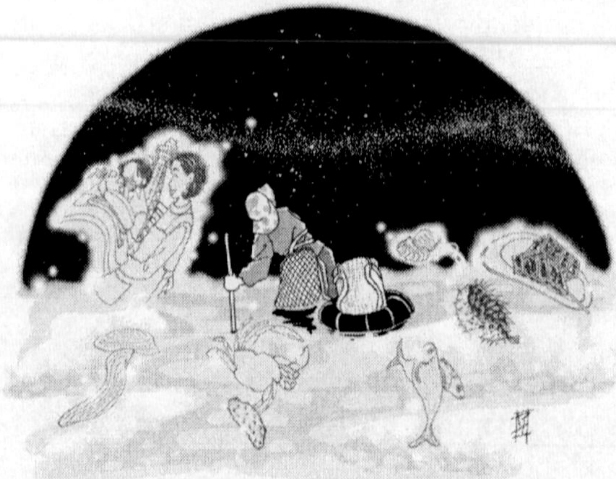

Figure 45: Night fishing in the tideland.
Source: Ikeguchi and Yanahara (2006)

The techniques used in girls' manga are very important here for expressing their memories and the sensing of signs in this scene. In this illustration, there is an old woman standing in the dark sea, around whom a variety of living things appear. Certain scenes from the past are revived, just like illusionary paintings. In black and white, even a fantastic illustration may face the risk of becoming a representation of some kind of horror scene in which a ghost wouldn't seem out of place. However, in this picture, the colours are gradated so that a touch of warmth may be produced. This way of fantastical representation in bicolour screen tone belongs to the techniques developed through manga for girls. Thus, this technique is used to illustrate what Grandma is remembering and how she is using her imagination, as well as the way she walks as if she is floating. With photos, it is not possible to illustrate fishing activities in the dark sea under a starry sky in Okinawa. What is even less possible is to show the content of her reminiscences, or the fish in the tideland as sensed by her. The use of actual photos can provide no more than a simplistic depiction of the scene or an introduction to the creatures in the tideland. What we would like to stress in this case is the accumulated memories and thoughts of the people who have engaged in fishing in the tideland. Also the fantastical feelings evoked during the activities diligently carried out by these people while they sense the smells and signs of fish in the blackness of the sea on a starry night. Unless you understand this, you cannot really expect to understand the values that people connected with the tideland have toward this sea. Moreover, the fact that the things created by the tideland bring about connections amongst these people cannot be understood merely in terms of economic values or numerical analysis.

This is the kind of thing that may only come to light when problems concerning welfare occur after the tideland is destroyed by development, which breaks up the relationships among local people. Also in this regard, it is considered that it is becoming increasingly important that the development of geography and the establishment of methodologies are promoted in such a way as to pursue a comprehensive understanding of the relations between people and the environment in a 'sensible' space.

In the historical process of developing such use of resources, locals learn from experience, and this knowledge is passed down from generation to generation. At the same time, they learn to develop their own intuitive perception, and to interpret the changing conditions and circumstances of the tideland when searching for fish. This happens certainly in the sensory world. Experiences accumulated by people engaged in fishing activities in the dark tideland are employed together with sensory/intuitive perceptions evoked whilst meticulously carrying out the activities. Locals are also sensitive to the smells and other signs of fish in the night sea.

The concept of 'Exciting Biogeography'

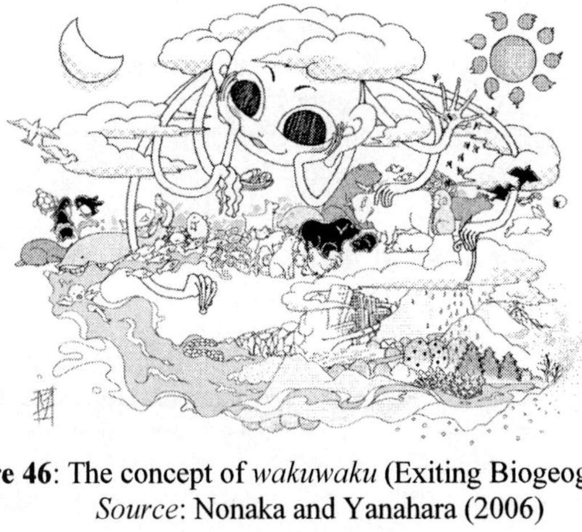

Figure 46: The concept of *wakuwaku* (Exiting Biogeography)
Source: Nonaka and Yanahara (2006)

The above figure was presented as a summary for the 'Exciting Biogeography' series, through which we attempted to describe the diversity and comprehensiveness of a living space where relations between people and the environment occur, and to show the significance of this kind of study.

In looking at the figure, the first thing that catches one's eye must surely be the wide-eyed expression of someone showing a great deal of interest in the living things. One will also notice the abnormally large ears. Their features suggest that the first thing to do in developing a relationship with the environment is to observe it and listen to it. Furthermore the person has many arms, with which they act upon and show affection for the living things. There may even be occasions where they are attacked by the environment. In spite of this, they still maintain their interest in observing the environment in much the same way as the characters appearing in the 'Exciting Biogeography' series. Moreover, the person in the figure is observing living things. The land, soil and climate across the bottom of the illustration are observed by the person through many layers of living things. We have used this figure to illustrate the points put forward in 'Exciting Biogeography' regarding the depiction of the relation of human feelings toward the environment. The physical environment is subject to topological, climatological and seasonal/periodic variations.

We propose that adequate consideration should be given to the fact that there is a great deal of variation in people's attitudes and behaviour towards living things inhabiting a given place. The following points were found to be significant in each of the case studies: peoples' attitudes towards living things and nature; the human practices resulting from these attitudes; the social, cultural, and historical aspects of these practices; and the natural environment on which they are based. All these elements are interconnected. The use of an illustration virtually guarantees an understanding of the above at a glance and by intuition. By looking at the details in more depth, or by raising questions and reasoning, one can develop a more comprehensive and intuitive/experiential understanding of the relationship between nature and people.

Conclusion: Towards a universal visualization of the sensory world

Until now, it has not been possible for geographers to clearly explain the relationship between people and the environment, despite its geographical significance. It is, however, demonstrated here that the use of the illustrative techniques of manga enables a pictorial visualization of the complex multi-layered world of reality, together with the sensory world, and the perceived relationship between these two worlds. People's activities and their feelings are set out in an illustration set against a background made up of a living world combining living things, nature, community, and culture. Perhaps this kind of visualization could be called 'geographical illustration'. This kind of geographical illustration presents the environment as something that is perceived subjectively by people, who create connections, and also as an aggregation of many elements interconnected on multiple levels, within which various actions and movements take place. Feelings evoked towards the existence of the environment and the actions upon it are sociological and historical constructions. Therefore, geographical illustration may embody the sensory world and show, on a two-dimensional level, that this sensory world exists in a relationship with many other things as a four-dimensional reality in space and time. Thus, people living in a place, and the thinking and values established by them in their world, is represented and objectified as a geographically integrated existence which is given meaning by, and connected to, all the contents and environmental elements represented in an illustration. Accordingly, as a geographical methodology, and also as a way of representing space, geographical illustration may not only contribute to the sphere of geography itself, but also make a more practical, universal contribution to society.

Spatial demonstration of discussions on relatedness

It is said that 'a picture says a thousand words'. With this in mind, one picture can enable readers to grasp by intuition the whole picture of elements and events. By looking at the shape, size, circumstances, and expression of elements constituting the picture, readers can have a continuous understanding of the relationship among the elements and their finer details. To perceive things starting from the whole picture and proceeding to the details is just the same as looking at a map to get information about a destination. Therefore, a geographical illustration may be considered a map of a discussion. By developing the concept of the conventional 'map' with its symbolization and precise scaling, one may expand its possibilities as a new system of spatial representation to be used for illustrating the effects of human activities and categorization, and the fact that the environment and people are dynamically inter-linked. This also indicates that one is not limited to the use of language as a means of explanation; visual representation is also a possible medium. In other words, the act of reading a paper can be replaced by the act of looking at an illustration. Illustration can be effective to help imagine ideas and values in modern society, where visual aids are widely used to attract readers' attention, and to make things easier to understand. In addition, the use of illustration to facilitate an intuitive understanding may make it easier to explain not only to experts such as geographers, but also to laymen, and particularly those who are not accustomed to academic study, thus being able to provide them with the basis for beginning a discussion.

Arousing empathy

A sociologically significant role of geography is to widely spread information regarding what is actually happening in the field, and so geographers are concerned with developing a means to do so. By representing peoples' feelings at the time they form relations with the environment. To express feelings is to move people, so this appeals to readers as an emotive force. As people share a common understanding of these feelings, readers can accept what an illustration represents, even though it may be unfamiliar to them. Thus it may easily arouse empathy. The utilization of manga may not only arouse interest amongst a wide variety of people but also promote a deeper understanding of different cultures and events that take place in faraway regions. Thus it may induce the reader to give serious consideration to the problems presented, compelling them to take action toward solving the problem. In the hope that this methodology may become more widely used in the future, we intend to continue studying the universality of both patterns of representation, and human emotions.

References

Ikeguchi, A. and Yanahara, N (2006) 'Aji to niwoi to asai umi (Taste, smell and the tideland)' In *Chiri* **51**(5): 68–71 (in Japanese)

K. Nonaka (ed.) (2004) *Yasei no navigation (Navigation in the wild-from ethnography to cognitive space science)* Tokyo: Kokon shoin. (in Japanese)

Natsume, H. (1997) *Manga wa naze omosiroika* (*Why is Manga interesting?)* Tokyo: NHK. (in Japanese)

Nonaka, K. (2005) *Ethno entomology* Tokyo: University of Tokyo Press. (in Japanese)

Nonaka, K. and Yanahara, N. (2005) 'Sasaretemo sukina hito (Loving Yellow Jacket)' In *Chiri* **50**(6): 17-20 (in Japanese)

Nonaka, K. and Yanahara, N. (2006) 'Owarini (Conclusion)' In *Chiri* **51**(5): 72–73 (in Japanese)

Nozomi, Y. (2005) 'Jikan resya de tabi wo suru (Take a time train for time travel)' In *LaLa-DX* **5**.

Park, K. and Nonaka, K. (2003) *Kankyo chirigaku no siza* (*Environment Geography—Perspective to Nature-human relationship study)* Kyoto: Showado. (in Japanese)

Simizu, I. (1999) *Manga Tanjo Taisho demokurasi kara no Syuppatsu (The Birth of Manga the beginning from the Taisyo democracy era)* Tokyo: Yosikawa, Kobunkan. (in Japanese)

Yanahara, N. (2006) '*Jikan resya de tabi wo suru* (Take a time train for time travel)' In *LaLa-DX* **5**

CHAPTER EIGHTEEN

PLACE MAKING IN THE CONTEXT OF URBAN DAILY MOBILITY PRACTICES: ACTUALISING TIME-SPACE MAPPING AS A USEFUL METHODOLOGICAL TOOL

PAOLA JIRÓN

Introduction

The everyday experience of place making in contemporary mobile urban societies requires an updated understanding of place as well as a way of representing this experience in a holistic manner. In simple terms, people identify places as those spots that they go to for some particular purpose or function, they also signify them and these in turn provide meaning to individuals and communities. However, the notion of place as signifying 'feeling at home' or as rooted or bound, settled and coherent communities has been questioned[1] due to the increasing pace of globalisation and time-space compressions[2] and for the fact that home is not always a positive or significant experience for everyone.

Thus places can best be understood as dynamic and mobile and not necessarily staying in one location.[3] As a large part of our days are spent going from one place to another, the practice of being mobile (or immobile) becomes essential in the consideration of these places and how they are constructed and experienced. Within this mobility, there are sporadic places that have special meaning to urban dwellers, but there are also everyday places, both of these can either expand people's experience of a city or confine it.

---

[1] Massey (1995) Cresswell (2001).

[2] Harvey (1990).

[3] Urry (2004).

Understanding the way people signify these daily encountered places as well as their meaning to urban practices requires innovative methodologies of enquiry and representation. Qualitative methodologies are usually used to explain urban practices through verbal description. However, visual and spatial forms of enquiry and representation are also required to adequately attempt to grasp them, though never fully, at least better. This paper proposes complementing time-space mapping, with an ethnographic research approach, as a comprehensive way of understanding mobile places, how they are created, and more importantly how they expand or restrict people experience of the city. This will be done using examples of three daily mobility experiences in Santiago de Chile.

Place making and mobility

Current globalisation processes, including technological, communicational and financial advances, as well as current planning, city marketing and branding trends, and global cities architecture tend to generate urban spaces that appear as clones of those elsewhere: suburban tracts, shopping malls, freeway interchanges, hotels, office complexes, rehabilitated old neighbourhoods, or gated communities are less and less distinguishable and often 'placeless' sites in Relph's[4] terms. Some believe that as spaces lose their distinctiveness, place loses its reality and significance; others insist that place persists as a constituent element of social life and historical change.[5]

According to Savage, place-making is still relevant today,[6] however, the process of place making in contemporary cities is complex. Massey argues that if social organisation of space is changing and disrupting the existing ideas about place, then the concept of place should be rethought altogether.[7] This is also because bound ideas of place representing identity and self value can exclude the migrant, the exiled, women, men, or children. In this view, places can be seen as the location of particular sets of intersecting social relations and intersecting activity spaces in time.[8]

Moreover, place making always involves an appropriation and transformation of space and nature that is inseparable from the reproduction and transformation of society in time and space. This is a progressive sense of place, where place is open and permeable and always in construction.

---

[4] Relph (1976).
[5] Gieryn (2000); Cresswell (2001); Sheller and Urry (2006).
[6] Savage *et al.* (2005).
[7] Massey (1994); Massey (1995).
[8] Massey (1995).

Furthermore places can have different scales,[9] types (private/public) or characteristics (fixed/mobile), all of which do not exclude each other and often overlap in time as well as in space. Thus space is not just a container of place but a disputed and integral part of it. As such, place is not only what is fleetingly observed on the landscape, a "locale" or setting for activity and social interaction, it is also what takes place ceaselessly, what contributes to history in a specific context through the creation and utilization of a physical setting.[10] This setting can be fixed, but it can also imply, and most often does particularly in contemporary urban life, mobility.

Place making is relevant today because these non-places, in Auge's[11] terms, are not simply un-rooted places marked by mobility and travel. Mobility and travel imply different forms of placement and displacement, by different genders, ages, religions, income groups, ethnic groups, etc. Places are about relationships, about the placing (or displacing) of people, materials, images, and the systems of difference they perform.[12] Place is both the context for practice as well as a product of practice, thus the practices that take place are of extreme relevance. As Cresswell mentions, places are never complete, finished or bounded, but are always becoming, in process.[13] Also, places are made through power relations which construct the rules which define boundaries. These boundaries are both social and spatial, they define who belongs to a place and who may be excluded, as well as the location or site of the experience. In fact, boundaries are in question and the incessant struggles cause these boundaries to be constantly shifting, thus creating the possibility of enlargement or confinement. In this sense, place is constituted through reiterative social practices, meaning that place is made and remade on a daily basis. This is what makes it especially relevant today.

In this context, mobility refers to all the ways people relate socially to change of place[14], which means more than the sum of journeys made. Mobility can be physical, but it can also be virtual or imaginative.[15] Urry[16] explains that travel has not been sufficiently researched except for the work of transport engineers and economists, who tend to examine simple

---

[9] McDowell (1999).
[10] Pred (1986).
[11] Auge (1995).
[12] Sheller and Urry (2006).
[13] Cresswell (2001).
[14] Bourdin (2003).
[15] Sheller and Urry (2006); Szerszynski and Urry (2006).
[16] Urry (2003b).

categories of travel, such as commuting for leisure or business. However, understanding such connections need not begin with the types and forms of transport, but with the experience of mobility, as the act of moving is not usually an end in itself but a means to achieve certain socially patterned activities. It is by untangling the experiences, that these practices and their consequences can be better understood, as would be the case, for instance, with the practices of going to work, taking children to school, or shopping. Understanding mobility from the individuals' practices, according to their spatial and socio-economic reality, broadens understanding mobility as transport, which can be limited to the supply and demand of infrastructure and means of transportation, and the number of journeys per person per day, according to motive, mode, itinerary and time.[17] Thus, while transport research neglects social processes involved in travel and how these affect people's lives, much of social science research has been a-mobile, ignoring or trivialising the movement of people for work and family, leisure and pleasure, and for politics and protest,[18] thus failing to examine how social life presupposes both the actual and the imagined movement of peoples from place to place, person to person, event to event. Mobilities are organised into complex patterns that transform the very social relations that the social sciences seek to explain.[19]

Furthermore, research has shown that mobility practices are hybrid. Therefore, what matters is to understand the situations of mobility and the experiences it gives rise to for the people involved and not just the finality of mobility. A comprehensive approach to the practices of mobility requires a grasp of the reality of contemporary movement but also an analysis of the genealogy of these practices.[20]

The notion of place in this mobile context is crucial, because, analysing mobilities involves examining the consequences for different peoples and places located in what could be called 'the fast and slow lanes' of social life. There is a proliferation of places, technologies, and 'gates' than enhance the mobility of some while reinforcing the immobilities of others.[21] Similarly there are certain experiences of mobility or immobility that enhance or restrict places. The time spent travelling is not the dead time that people always seek to minimise; activities occur while

[17] Montezuma (2003).
[18] Sheller and Urry (2006).
[19] Urry (2003a).
[20] Lussault (2003).
[21] Sheller and Urry (2006).

on the move, therefore, travel is not just a question of getting to a destination.[22]

A key issue about incorporating the analysis of urban daily mobility practices to understand place making, is that these practices may actually create place confinement or place enlargement. In the analysis of exclusion and social segregation, the work of Pierre Bourdieu in *The Weight of the World*[23] is crucial, as it explores the social suffering in contemporary society marked by neo-liberalism, the dissolution of class identity and the retreat of the State. One specific form of *misere de position* is related to the experience of occupying a low/marginal position in a social and geographical hierarchical space. While people who possess financial, cultural and social capital have power over space and tend to self-segregate, people with scarce resources are confined to places they do not choose. Bourdieu argues that in the context of the de-industrialised society, the former working-class communities have suffered a symbolic transformation into places of social deportation, where people find nothing in common but the lack of economic, social and cultural capital.[24]

This place confinement appears relevant to incorporate in a study of urban daily mobility, as it can be understood as the restriction or limitation of individuals to place making in mobile contexts due to physical, social, economic, cultural or other forms of boundaries. Thus the need to clarify whether, despite the high levels of mobility in urban areas, the point of departure contributes to perpetuate the existing inequality, thus confining places along the daily journeys or, whether regardless of the point of departure, places are expanded or enlarged in the experience of being mobile in the city. In this analysis, the point of departure, i.e., housing, socioeconomic conditions, past history, habitus in general, may contribute to place confinement/enlargement, or perhaps the actual experience of urban daily mobility, plays a more important role in urban place making.

The above discussion promotes the idea of understanding that the process of place making in the context of urban daily mobility requires rethinking both the object of enquiry as well as the methodologies for research, this is mainly because there is currently a difficulty in terms of explanation and representation of the implications of daily mobility in the social sciences. Urban practices require more detailed explanations, however, although lived space can be mapped, it can never be captured by conventional cartographies, it can be creatively imagined as it obtains

---

[22] Sheller and Urry (2006).
[23] Bourdieu (1999).
[24] Bourdieu and Wacquant (1992); Bourdieu (1999); Castro and Lindbladh (2004).

meaning only when practiced and fully lived.[25] Through an ethnographic approach to researching urban daily mobility, the following section explains the way time-space mapping can be used today in urban daily mobility research as a form of representing the experience of mobility. This will be exemplified through preliminary results using maps, photography and narratives, in the aim of getting a step closer to understanding the lived space of urban mobile experiences.

Ethnographic approach to urban daily mobility using time-space mapping

For this research, an ethnographic approach was chosen as the best way to describe the everyday mobile lives of individuals, given its flexibility and possibility of penetrating, through deep description, into the daily routines of mobile urban dwellers. With the aim of analysing urban daily mobility practices in Santiago de Chile, the research was undertaken in a delimited, mixed income sector on the South-eastern part of the capital. A three stage process was undertaken with selected households: the first stage involved in depth interviews regarding the person's history, background, choice of current place of residence, and a detailed description of a regular day. This daily routine was explained using maps and a 24-hour diary. The second stage involved accompanying each person on a regular weekday, from the time they left the house until their days were over. Details of the journeys were recorded as well as photographed by the researcher and the interviewees. The final stage involved further interviewing on issues not covered during the previous stages particularly in terms of access and place making. The information gathered is currently being processed to reach the research objectives through the analysis of time-space mapping as well as narratives and photography.

Time-space mapping is part of a disciplinary area within geography denominated time geography. In its integral view of the world, time geography attempts to bring together the different elements of life experienced through a unifying lens. The approach was originally formulated during the 1970s by Swedish geographer Torsten Hägerstrand, and it was initially used mainly in terms of transport planning. During the 80s and 90s it was also applied in a diversity of fields from city planning to social equity. Hägerstrand's visionary work implied an attempt to bring together knowledge from distinct scientific areas and from everyday practice and to

[25] Soja (1999).

reveal relations taking place in this milieu.[26] Through his work, he implied that the study of aggregate populations masked the true nature of human patters of movement, believing that an understanding of disaggregate spatial behaviours was paramount.[27]

Within this approach, Hägerstrand developed time-space mapping to illustrate how a person simultaneously navigates his or her way through the spatial temporal environment[28] (see figure 47). He used this notation to demonstrate how human spatial activity is often governed by limitations and not by independent decisions of spatially or temporally autonomous individuals. He identified three categories of limitations or constraints: capability constraints (due to physical or biological factors); coupling constraints (referring to the need to be in one particular place for a given length of time, often in interaction with other people); and authority constraints (an area that is controlled by certain people or institutions that set limits on the access to particular individual or groups).[29]

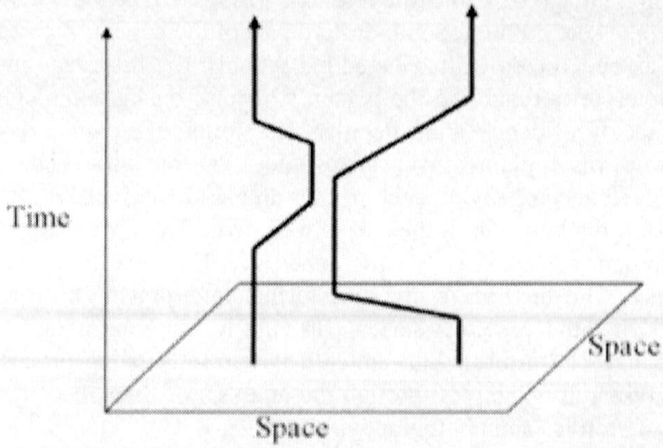

Figure 47: Time-Space map

There were some criticisms to the approach as it was seen as a mechanistic device for social engineering.[30] Also, during the 80s, though recognising all the possible attributes, Giddens considered the approach as naïve in

[26] Hägerstrand (1970).
[27] Corbert (2005).
[28] Corbert (2005).
[29] Corbett (2005).
[30] Lenntorp (1999).

treating individuals as coming into being independently of the social settings they confront in their day-to-day lives. Also, he expressed that it perpetuated the dualism of actions and structures, giving little attention to the essential transformation character of all human actions. Thirdly, he criticised that the emphasis of people through time and space was seen only in terms of constraints and limitations, without considering the possibilities which may arise in movement.[31] Finally, Giddens criticised the weak developed discussion on power, invoking a zero sum conception of power as a source of limitation upon action, where little possibility for generative power exists.[32]

Over the past few years, there has been a revision or revival of time-space mapping as a useful tool in various disciplines. It is still considered powerful in terms of providing a sense of concreteness; as seeing that space and time are not simple social categories but actual constraints on human action; as providing a geographical ethics in terms of the wise use of time and space; as providing a language to explain time and space; and most importantly, in terms of its visual representation in maps and diagrams.[33] Today, the approach is being incorporated in various areas, from travel behaviour in transport planning using origin and destination surveys,[34] spatial mapping using GIS or eventual virtual interaction[35] or the human activities effect on the natural environment.[36] It has also been used in gender studies[37] or migration analysis,[38] all of which have a quantitative approach to human behaviour. Other ways to see the potential of time-space mapping can be taken further as in the use in art[39] or artistic performances.

There have also been attempts to understand urban daily mobility in qualitative terms. Latham[40] suggests the diary-photograph/diary interview method, where people are solicited to write diaries and photograph their daily experiences and photograph the interesting and/or significant places and events of their week. Though innovative in the way of inviting interviewees to participate in the research production and interpretation

---

[31] Giddens (1985).
[32] Giddens (1985).
[33] Thrift (1995).
[34] Newsome *et al*. (1998).
[35] Miller (2005).
[36] Peuquet (1994).
[37] Kwan (2002).
[38] Southall and White (2005).
[39] Thrift (2005).
[40] Latham (2003); Latham (2004).

and thus minimising the input of the researcher, there are difficulties in convincing participants to fully engage in the process, particularly when their daily mobility experience is filled with time limitations and multitasking to make end meets, thus risking being able to obtain accurate, detailed or any information at all. For the aim of understanding the experience of mobility, mobile place making and the generation of inequality through mobile practices, this research applied an ethnographic account of the journey, where, after a period dedicated to obtaining access and generating trust among 12 families living in a specific area of Santiago de Chile, the researcher joined the research participants in their daily journeys. These journeys were complemented with photography, daily activities charts, interviews, mental maps, and in some cases group discussions which provide more detailed descriptions of the experiences and the possibility of triangulating the information. Using time-space mapping as a way of representing these experiences, photography and narrative greatly contribute to understanding the complexity of urban daily mobile routines in the city of Santiago, and their social, economic, cultural and physical implications.

It should be said, that through these approaches, the whole picture is never wholly revealed, nor is it their intention, but this implies a need to work towards creating a better methodologies to capture mobility experiences, in terms of its production, as in their interpretation and representation. This requires a variety of methodological instances such as observation, interviewing, accompanying travellers in their daily journeys, as well as aggregate travel patterns, but also moving towards creating methodologies to generate greater participation in the research process from those involved. Such endeavours may require longer time for fieldwork, particularly in terms of building trust relationships.

Mobility in Santiago de Chile: Journeys of Bernardo, Marta and Carlos

> El mundo pasa allá en otra parte mientras yo me quedo en la casa
> The world is happening elsewhere while I stay at home
> (Alicia, housewife)

Using analytical narratives, photographs and time-space mapping to represent daily mobility activities, the following section describes the journeys of three people living in Santiago de Chile. These are the mobility experiences of Bernardo, Marta and Carlos…

Bernardo is 40 years old, married to Alicia and they have three children. He has lived in Santiago all his life. He finished high school and by the age of 18, Alicia became pregnant and they got married. Life during the first years of marriage was difficult in economic terms, living at their parents' homes, making very little money and 'raspando la olla' (scraping the pot), literally he says, "it was tragicomic, but we really didn't make enough to eat". He says they lived like gypsies for years, from one place to the next, with their relatives in overcrowded spaces. It took them many years to get the house they live at now, through the government application system. They have been living at their current house for about 6 years. His job has also improved and now works in a printing shop; he's been working there for almost 8 years now and enjoys it.

Looking at figure 48 below, Bernardo's daily routine starts at 6:00 when he gets up. He leaves his house at 6:30 in order to catch the bus from the bus terminal close to his house and avoid traffic as much as possible so he can make it comfortably in time for his job, which starts at 8:00 am. His journey to work is an enjoyable one, he says. He's been taking the same route for many years, so although the other passengers are not his friends, he knows many of them and, since he enjoys any available possibility for conversation, he always ends up talking to the person sitting next to him. He doesn't like sleeping on the bus, unlike most of the morning riders; he's scared he might fall off his seat. He doesn't like looking out the window either, he knows the scenery by now and it doesn't impress him much, he says. What is more interesting for him is what happens inside the bus. 2) A few streets away from his neighbourhood, his friend Angela gets on the bus. They talk for the whole ride, about their friends, their family, their jobs, other passengers, life… 3) She says Bernardo prefers sitting next to pretty girls and chatting them up. He just smiles. He nods and waves too many passengers on the way. 4) He helps pass money to the driver from the back passengers who hop on the bus from the back entrance, he then helps to pass the ticket and change back. 5) When he arrives at his stop, he walks Angela to her job at a factory, then walks to a bakery to buy bread and ham for breakfast. He walks to work and waits for his boss to arrive and open the door. He stays indoors all day, gets his lunch delivered to work.

6) At 6:00 pm he showers and leaves his work. He walks to the main street to catch his bus. The neighbourhood is known to be dangerous, but he knows most of the people who go by. 7) The bus going home is full, so he hops on it from the back. He meets some people on the bus. He's on the

lookout for thieves who usually get on the bus on the 15ths and 30ths of the month, when people get paid and the buses are full. He's had his phone stolen before. He says people usually signal to each other when a 'lanza' (thief) gets on. But no one does anything. 8) He makes it home early on weekdays. His neighbourhood is buzzing with music, laughter, screams, kids playing and people on the street. Inside his house, the noise from outside makes being indoors unbearable, he says it's easier being outside…

He says he wouldn't change his job to one closer to home as he actually enjoys taking the bus, talking to strangers and people in general. He doesn't mind the daily hour and half on the bus, since he doesn't see it as dead time but as a way of socialising and relaxing. Life for him is about having fun, enjoying himself; he loves singing, playing guitar, socialising. On Fridays he takes his guitar to work and goes out to pubs and parties to play, usually returning home the next day. He doesn't want much more in life. He thinks people should be friendlier, sees how incarcerated people are in their houses, how few opportunities they have to see the world, even by going to the park, he knows people can open their minds and think differently and not feel caged in. He knows there are inequalities in the city, and that there are parts of it that because of money, access or simply because of his clothes or looks he doesn't fit in, so he just won't go there.

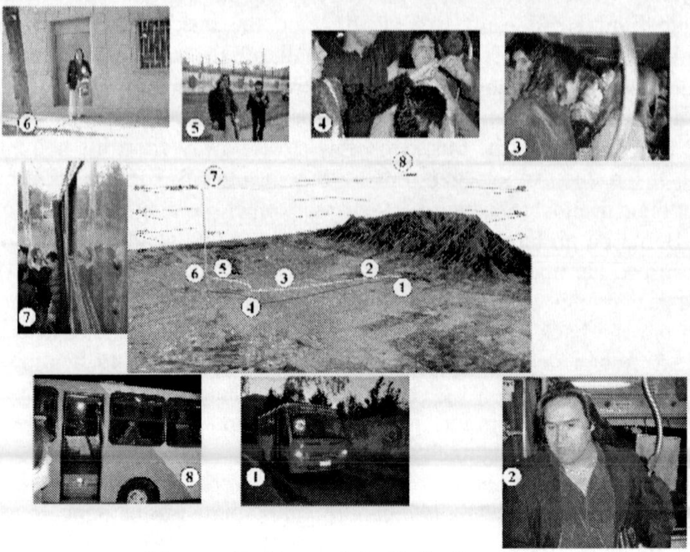

Figure 48: Bernardo's Daily Routine.

Marta is 24 years old and is married to Luis. They have a 7-year-old daughter called Patricia. They both come from a small farmers' town in the North of the country. Because of family problems she was sent off to live in foster homes since she was very young. By the time she finished 7th grade, she decided she didn't want to go to school anymore. At the age of 17 she moved to Santiago, to work as a house cleaner. She went back to her town a few years later, got pregnant and married Luis. She didn't like the country life style, it bored her, so when Patricia was about 3 years old she went back to Santiago to work as a cleaner and Luis' mother took care of her. She'd go back to see her once a month, for four days at a time, but it was difficult she says, Patricia wouldn't recognise her at the beginning. She thinks that it's because of this period that she's now believes that her mother doesn't love her enough and is not going to come back. After a few years of this, her husband decided to move to Santiago, and they rented a room downtown. She applied for a house through the government housing system and managed to obtain the house they are living in now. She now works cleaning in a rich neighbourhood in the North-east of Santiago, but it takes her almost two hours to get there. She's thinking of quitting because of the distance and time she wastes moving back and forth, but mainly because she doesn't want her daughter to be alone and she doesn't have anyone to pick her up from school.

Martha (see figure 49 below) works three days a week as well as every second weekend. She gets up at 6:45 to prepare Patricia for school. By 7:30 they leave the house and walk through the park, an open wasteland, and a bridge, to get to school. She drops her off, waits until she's let in and walks to take the bus. She walks further than needed because at this time the buses are full and if she waits by her stop, she won't get let in. 2) The sidewalks are not paved and the air is very dusty. She manages to get on the bus and push her way in. 3) Effectively, when the bus gets to her stop, the driver doesn't even bother stopping. The ride lasts about an hour and slowly she moves towards the middle of the bus. The bus is so full, it's almost impossible to see outside. It's mainly women riding on it, most of them going to the richer side of town to work as house cleaners. By the time she almost reaches her first stop, she manages to get a seat. 4) She gets off and quickly walks to the corner as she sees her next bus approaching. This one is also full and she gets on it from the back. It is also full of women and some male construction workers. The music is very loud and she likes it. This ride takes about half an hour and she manages to get a seat almost one block away from her final destination, she's the last one off the bus. 5) She then walks through beautifully gardened houses, with pretty dogs and nannies wearing uniforms to her job inside a gated community. She doesn't leave the house all day and finishes around 7:00 pm. She waits for the bus for over 20 minutes and manages to grab a seat. 6) She takes her second bus and manages a seat as well. It's dark when the bus gets to her final stop. She buys tomatoes, bread and butter in a corner shop on her way home and walks for about 20 minutes home through the same route she took in the morning. 7) It's very dark but she knows her way. When she gets home, Patricia is at home alone, watching TV while she waits for her mother. She has been alone for about an hour, has done the washing up and had her tea.

She doesn't find anything about her journey interesting or worth remembering. She can't sleep on the way there because she stands and can't read on the way back because she gets dizzy and is too tired. She doesn't remember the route or know the names of the streets although she has been making the same journey for over two years, she's oblivious to her surroundings. The journey is just a necessary step to get to work.

Her boss has asked her repeatedly to get up earlier so she can get to work earlier, but she has explained that she can't get there any earlier because she has to take her daughter to school and then the buses are overcrowded. The boss says it's not her problem; she just needs to get there earlier. So she decided to tell her boss she was going to give up the job, and the boss told her to give her enough time until she finds a replacement. Marta is therefore looking for a job closer to her place. She likes having days off but then she gets absolutely bored, she gains weight and feels trapped, she says. But she doesn't know what to do; she doesn't have anyone to leave her daughter with and can't afford to pay for it. She doesn't want her daughter to have the life she did.

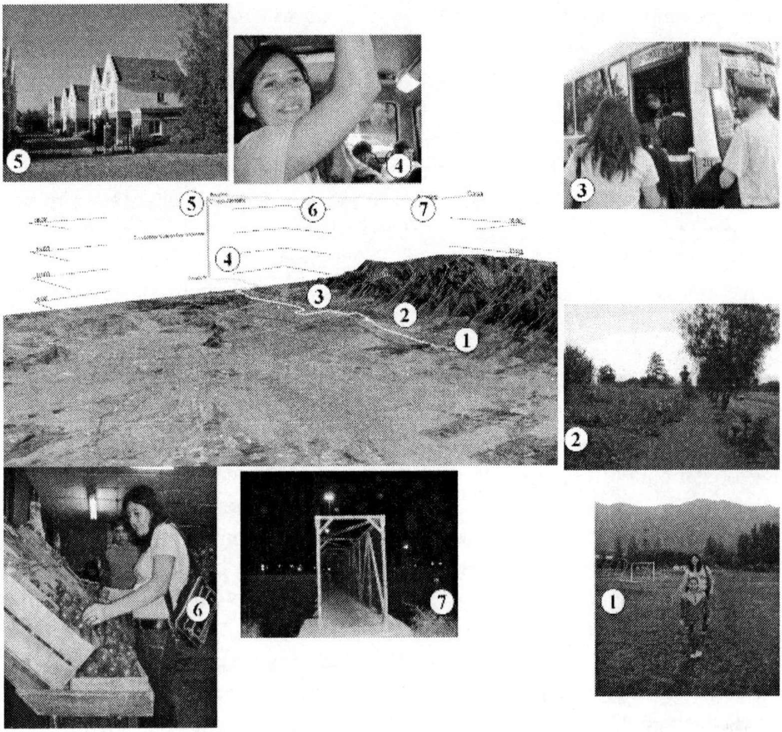

Figure 49: Marta's Daily Routine.

Carlos is 50 years old, comes from the South of Chile, is married Laura and they have three children. He studied to become a teacher and practiced as such for many years. About 5 years ago, he decided it was more profitable to manufacture and sell school uniforms to private schools. He started small, making sweatshirts for the school he worked at, but he now provides uniforms for at least 6 different schools, and even manages to make special orders for out of town clients. His job is informal, he buys the material himself, cuts some of the material at home, in his living room, but most of the sawing, embroidery, cutting and pasting, takes place in informal workshops throughout the city. He has a room in his house where some of the uniforms are stored and displayed, but most of the selling occurs in the school's parking lot when he picks up his kids. He doesn't have a 'regular' routine, since, apart from trying to make his business work, he manages various aspects of home keeping as well as his wife's extra income generating activities. According to Carlos he is the "family's driver, secretary, busboy, nanny..."

Carlos' daily journeys starts at 6:30 in the morning when he quickly gets dressed and drops his wife a few blocks away from home where she's picked up to go to work (see figure 50). He then goes home 1) to get the children ready for school. He makes sure they clean up, have breakfast, have their bags ready and takes them to school. He takes alternative routes to avoid traffic and stop lights. 2) Once there, he opens the car boot and people who already know him, approach him to buy. Once the parents leave, he packs up and heads to the petrol station, to pump up for the day. He stops to talk to his wife and receive the instructions for the day. By 8:30 the day ahead seems tough. He goes to the shopping mall to park his car and starts his errands. Banks open at 9:00, so he has to wait. 3) From 9:00 to 10:00 he manages to go to two banks, pay the water, electricity, phone and gas bills at each service, they are all quite close to each other. He doesn't have the invoices, so it takes him longer because he has to queue up to get them. By 10:30 he goes to the supermarket to buy bread for breakfast, he rings the cleaning lady who is at home by now, she tells him what to buy for lunch. Back at home he makes himself breakfast and by 11:30, he has to rush back to school, because the children are finishing early today. He picks them up and opens the car boot to sell again. This is the best time of the year for him, since school just started, the parents are buying the yearly supply of clothes: trousers, shirts, aprons, sweatpants, sweatshirts, t-shirts, jackets... 4) After dropping the kids at home, he heads off to Independencia, an area in down town Santiago which specialises in fabrics and various materials. Though he has his regular shops, he shops around for the exact colour and the best prices. After an hour and half of visiting at least 20 shops, he goes to the embroidery workshop, not far away to leave some material, it's closed so he goes to another shop to pick up some pockets but it is also closed. So he decides to have lunch at a diner close by.

By 2:30 he returns to the two previous shops and his wife rings to give him more instructions for the afternoon. 5) He goes to Laura's business, to pick up some boxes on the way to her job at the hospital where he picks her up and they set off to the south of the city, to her third job. He then stops by at home to drop off some material and pick up clothes. Soon after, he goes to one of the workshops to pick up cotton shirts. 6) On the way there, he stops by the supermarket to ask for cardboard boxes. Once at the workshop, he realises that no one is home, so he goes back home to get the keys to the house. Once at the workshop again, he packages the shirts, in individual bags first and then in a box. This package is for a client who is travelling to the South and needs to make the delivery by 5:00 today. 7) After delivering the packages, he stops by another workshop to pick up badges. They are not ready, so he has wasted the trip. He stops by the supermarket again to buy things for the kids to take to school. As he heads back home his wife calls again. She's ready, so he picks her up. The day is almost over; they stop by on the way home to buy bread. They arrive home shattered, but it's time to prepare the children for the next day. They don't sit down for dinner as the table is full of cutting machines; the children make their own sandwiches, while the parents mechanically operate. Tomorrow should be a calmer day he says...

He doesn't remember much from the journey, he doesn't really notice what is happening outside, as he's mainly concentrated on the driving, on answering the phone, and on organising all the things he needs to do in his head. The moments that are important to him and perhaps significant are those he spends in the car with his wife. They are the only time of the day where they manage to talk, to discuss the kids, to tell each other about their days, their worries and concerns. It's their time to be together. At times, when he has to go out late at night to pick up clothes or drop off material, she goes with him, just for the ride, just for the chance to accompany him, to be with him and spend time together. Their days are so packed with information, that the time in the car is their private time, the car itself is their place, their significant place where they decide upon their lives.

This incessant moving is just the way he assumes life is, constantly performing different activities; it's the life he's choosing, for his children to be better, to be able to live in the house they bought. They represent Chilean middle income group with aspirations to improve in life and give a better life and more opportunities to his children. He worries that he won't be able to make enough money, as they depend on his wife's permanent and considerably higher income to make things work. He stresses about not being able to keep up, about life going by without him noticing, while he's driving. He fell very ill last year and he didn't have enough medical insurance to cover for it. He was very stressed, about having just bought the house and not being able to pay. His role as a family provider is changing and his wife is assuming that role. He doesn't mind it, but he says it stresses him out, even if he doesn't show it.

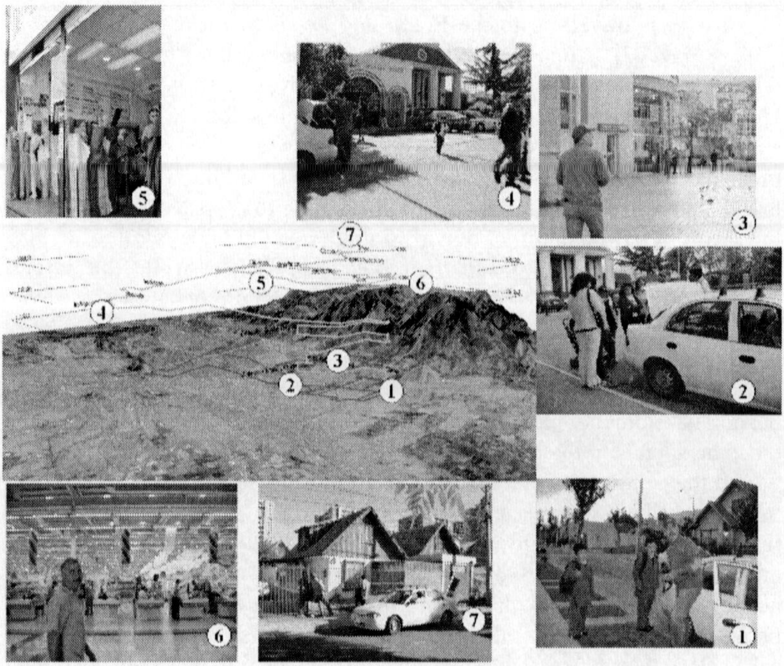

Figure 50: Carlos's Daily Routine.

The complexity of differentiated mobility

The above three stories provide a brief description of the way daily mobile routines in Santiago de Chile might be like. As can be seen, they shed light on mobile practices that then can illuminate different aspects of contemporary urban living. They describe representations which illustrate the possibility of unveiling the complexity of differentiated mobilities.

An initial complexity relates to gender differences and their implications on a daily trajectory and development of daily activities. Such is the case of Bernardo and Marta who belong to a similar income group and live in the same housing complex; however, their experiences differ. The first is more comfortable, because Bernardo can leave early in the morning, thus enjoying the experience, while Marta has to leave later and endure the overcrowded bus while standing. The reason why Bernardo can leave earlier, although his boy is the same age as Marta's little girl, is that his wife Alicia gets his child ready for school, whereas in Marta's case, she's the one in charge of this task while her husband works. Marta's concerns oscillate between her daughter and her job, she knows there is no way she can stay at home as she gets bored and the money she makes helps at home, but she says she needs to be there for her daughter, unlike her parents, who sent her to a children's home when she was very young. Four hours daily on the bus are long and difficult, but regardless of getting a job close by, she still needs to find someone to pick Patricia up from school. Gender differences, the way traditional household roles perpetuate, have relevant implications in daily lives, including the way people experience mobility.

Income also generates a difference, as is seen in the case of Carlos. This experience, though different than the previous ones, due to the use of the car, is nonetheless difficult, intense and tiresome. Having a mobile and flexible job requires fast and available mobility, which not only burdens the worker, but often involves the whole family. In these circumstances, the city itself does not make the routines easier and the car is a necessary means to accomplish all the tasks. Daily life for middle income groups in Santiago in increasingly difficult, attempting to juggle the multiple needs of the family with limited resources generates stress, an unavoidable symptom of modern times. This family, as well as others interviewed, recognise the daily difficulties to access activities, relations and places, it's "a machine that doesn't stop" and "all of a sudden, the year is over". They recognise that they have immersed themselves in a cycle of debt to reach what they aspire for their children, not for themselves, and this cycle is clearly reflected in their daily lives.

A more detailed look at the experiences of the different members of these families, reveal different ways of place making, as is the case of the youth, who in their access to leisure, generate new places, new relations and new practices, purposely skipping places that confine them. Other experiences, such as the elderly, struggle to find autonomy through mobile place making, or the single mothers who endeavour to develop friendships that help to cope with their daily routines, and enjoy the places encountered daily. Other experiences reflect the inequities present in the city of Santiago, in access to activities like education, employment, health, but also leisure and relations. The inequities revealed, bring about a new discussion in the way urban segregation in Latin American cities is being approached, and how through urban daily mobility analysis, new responses to urban inequality can be addresses. Finally, the experiences reflect the inadequacy of current urban planning policies and interventions, particularly housing, infrastructure, urban design. By providing a closer look at the way people are living in cities today, these formal interventions could have a better chance at improving urban quality of life.

As can be seen time-space mapping, complemented with an ethnographic account, can be useful as a way of tracking mobility in order to understand the way people move about the city, comparing movement, while using the same basic information. It can also help to visualise movement as a way of dimensioning the extent of mobility (or immobility) along the city. It can help to compare different trajectories, and the time and space used. But more importantly, it can be a complementary way of providing more depth to a description of a situation, especially if combined with an ethnographic approach to understanding urban daily mobility. This methodological approach can be useful in contemporary urban research as it provides a whole range of possibilities for representing information that would otherwise be lost in the research process.

References

Auge, M. (1995) *Non-Places. Introduction to an Anthropology of Supermodernity*. London: Verso.

Bourdieu, P. (1999) *The Weight of the World: Social Suffering in Contemporary Society*. Cambridge: Polity Press.

Bourdieu, P. and Wacquant, L. (1992) *An Invitation to Reflexive Sociology*. Cambridge: Polity Press.

Bourdin, A. (2003) Workshop 4: Hypermodern individuals? International Colloquium "The sense of movement. Modernity and mobilities in contemporary urban societies" organised by IVM, Centre culturel international de Cerisy-la-Salle, Paris, France, Institute pour la ville en mouvement.

Castro, P. B. and Lindbladh, E. (2004) 'Place, discourse and vulnerability - a qualitative study of young adults living in a Swedish urban poverty zone' In Health and Place **10**: 259–272.

Corbett, J. (2005) *Torsten Hägerstrand: Time Geography* CSISS Classics.

Cresswell, T. (2001) 'Introduction: Theorizing Place' In G. Verstraet and T. Cresswell (eds.) Intersecting: Place, Sex and Race. Mobilizing Place, Placing Mobility: The Politics of Representation in a Globalized World Amsterdam: Rodopi. (11–32)

Giddens, A. (1985) 'Time, Space and Regionalisation' In D. Gregory and J. Urry (eds.) Social Relations and Spatial Structures' London: Macmillan Publishers Ltd.

Gieryn, T. F. (2000) 'A Space for Place in Sociology' In Annual Review of Sociology **26**: 463–496.

Hägerstrand, T. (1970) 'What about people in regional science? In Papers in Regional Science **24**(1): 7–21.
Harvey, D. (1990) *The Condition of Postmodernity: An Enquiry into the Origins of Cultural Change*. Oxford: Blackwell Publishers Ltd.
Kwan, M.-P. (2002) 'Feminist Visualization: Re-envisioning GIS as a Method in Feminist Geographic Research' In Annals of the Association of American Geographers **92**(4): 645–661.
Latham, A. (2003) 'Research, performance, and doing human geography: Some reflections on the diary-photograph, diary-interview method' In Environment and Planning A **35**(11): 1993–2017.
—. (2004) 'Researching and Writing Everyday Accounts of the City. An introduction to the diary-photo diary-interview method' In C. Knowles and P. Sweetman (eds.) Picturing the Social Landscape: Visual Methods and the Sociological Imagination London: Routledge.
Lenntorp, B. (1999) 'Time-geography – at the end of its beginning' In Geojournal **48**: 155–158.
Lussault, M. (2003) Workshop 3: Mobilities elsewhere. Comparative approaches. International Colloquium "The sense of movement. Modernity and mobilities in contemporary urban societies" organised by IVM, Centre culturel international de Cerisy-la-Salle, Paris, France, Institute pour la ville en mouvement.
Massey, D. (1994) *Space, Place and Gender*. Cambridge: Polity Press.
—. (1995) 'The Conceptualization of Place' In D. Massey and P. Jess (eds.) A Place in the World? Places, Cultures and Globalization. New York: Oxford University Press and The Open University.
McDowell, L. (1999) *Gender, Identity and Place. Understanding Feminist Geographies*. Cambridge: Polity Press.
Miller, H. J. (2005) 'A Measurement Theory for Time Geography' In Geographical Analysis **37**: 17–45.
Montezuma, R. (2003) ' Ciudad y Transporte. La movilidad urbana ' In M. Balbo, R. Jordan and D. Simioni (eds.) La movilidad urbana. La Ciudad Inclusiva Santiago: CEPAL **88**: 175–191.
Newsome, T. H. et al. (1998) 'Urban activity spaces: Illustrations and application of a conceptual model for integrating the time-space dimensions' In Transportation **25**: 357–377.
Peuquet, D. J. (1994) 'It's about time: a conceptual framework for the representation of temporal dynamics in geographic information systems' In Annals of the Association of American Geographers **84**(3): 441–461.

Pred, A. (1986) *Place, Practice and Structure. Social and Spatial Transformation in Southern Sweden: 1750–1850*. Cambridge: Polity Press.

Relph, E. (1976) *Place and Placelessness*. London: Pion.

Savage, M., Bagnall, G. et al. (2005) *Globalisation and Belonging*. London: Sage.

Sheller, M. and Urry, J. (2006) 'The new mobilities paradigm' In Environment and Planning A **38**: 207–226.

Soja, E. (1999) *Postmodern Geographies: The Reassertion of Space in Critical Social Theory*. London: Verso.

Southall, H. and White, B. (2005) *Mapping the Life Course: Visualising Migrations, Transitions & Trajectories*. London: Lifeline project.

Szerszynski, B. and Urry, J. (2006) 'Visuality, mobility and the cosmopolitan: inhabiting the world from afar' In The British Journal of Sociology **57**(1): 113-131.

Thrift, N. (2005) 'Torsten Hägerstrand and social theory' In Progress in Human Geography **29**(3): 337–340.

Urry, J. (2003a) 'Social networks, travel and talk' In British Journal of Sociology **54**(2): 155–175.

—. (2003b) Workshop 1: Beyond societies. International Colloquium "The sense of movement. Modernity and mobilities in contemporary urban societies" organised by IVM, Centre culturel international de Cerisy-la-Salle, Paris, France, Institute pour la ville en mouvement.

—. (2004) 'Death in Venice' In M. Sheller and J. Urry (eds.) Tourism Mobilities: Places to play, places in play. London: Routledge. (205–215)

CHAPTER NINETEEN

THE ETHICAL AND AESTHETIC VALUE OF "AUTONOMIC" NATURE FOR OUR ETHICAL AND AESTHETIC ATTITUDES TOWARDS THE ENVIRONMENT

CHRISTIAN W. DENKER

Introduction

Why is the aesthetic appreciation of autonomous nature important for our ethical and moral attitudes towards the environment? I will reflect on this question in three steps that are represented by the three parts of my paper. The first part addresses the relation between art and nature. I will point out possible differences and similarities between models for correct aesthetic appreciation for art and for nature. I will argue that there are no definite conditions for the aesthetic appreciation of nature neither for the aesthetic appreciation of art. The relativity of aesthetic pleasure will be discussed in an argument on imagination and autonomous nature. The second part explains the possible impact of autonomous nature on ethic theory. In this context I will first reflect on possibilities to link aesthetic and ethic theory, then on emotional connection between aesthetic and moral values and lastly on aesthetic pleasure as a part of good life. In the last part I will elaborate the impact of aesthetic pleasure on a good life and conclude on the specific interest of aesthetic pleasure for our moral behaviour towards nature. My main interest is to show, that our attitude towards art and towards nature depend one on each other and that they are both helpful elements for the definition of moral behaviour towards the global environment and towards our personal life.

The aesthetic appreciation of nature and of art

Why is the ideal of autonomic nature important for our ethical and aesthetic attitudes towards the environment? Whenever art is practised, nature seems to lose its autonomy to human activity. The relationship between art and nature however, is not necessarily opposed. In fact, both are part of our environment in the sense of the German term *Lebenswelt* as it has been used by Husserl.[1] The similarities and differences in the aesthetic appreciation of art and the aesthetic appreciation of nature support this assumption. The question of how autonomic or independent nature is formed by artificial creativity arose out of the increasingly generalized understanding that artificial environments can be destructive towards human life. Especially after the 20th centuries' ecological turn of environmental philosophy, problems such as pollution, the extinction of species, and the destruction of wild life emerged as criticisms to human centred ethics. In this context, the ideal of autonomic nature has inspired the revision of ethical rules for human interaction with the environment and how art could help improve these rules.

Does the aesthetic appreciation of art or of nature help us to define standards for our moral behaviour towards the environment? In order to understand the significance of this question, we might remind, that the increasing importance of artificial creativity for the development of our personal environment is a reference point for behavioural rules towards nature. In fact, modernity reversed the traditional understanding of art as representation. Since the autonomy of art has been established,[2] nature's superiority over art is not an uncontroversial truth anymore. With his famous thesis that artistic beauty is a product of human spirit and therefore superior to any natural beauty, Hegel set a historical marking point for human interaction with nature.[3] Since then, philosophy has rarely focused on nature to explain the structure of aesthetic values. Instead, when trying to explain our aesthetic attention to nature, art has become the focal point and model for aesthetic appreciation of any kind. Therefore it seems that only art can help us to define rules for our behaviour towards the environment. I do not agree with this. The main interest of my following thoughts will be to underline the ethic and the moral importance of aesthetic appreciation of nature.

---

[1] Husserl (1993: 50).
[2] Kant (1992: 243–246).
[3] Hegel (1970).

Models for correct aesthetic appreciation

In order to explain the importance of aesthetic appreciation of nature for our behaviour I will first reflect on differences between the aesthetic appreciation of nature and the aesthetic appreciation of art. Works of art can be considered as a group of objects that differ from other objects in a specific way. For example, objects of art can be defined in relation to their creator, by claiming their symbolic value or simply their social function. However, objects of nature that satisfy our aesthetic sensibilities can't be distinguished in this way from other natural objects that do allow an aesthetic experience. Hence, it can be argued that objects of art are also dependent on aesthetic appreciation. The distinction lies in that the aesthetic appreciation of art depends on specific objects, while the aesthetic appreciation of nature does not necessarily. To clarify: the conceptual object of aesthetic nature is based on our aesthetic experience. In contrast, it is only in art that our aesthetic appreciation is based on a specific group of objects explicitly identified.[4]

Are there definite rules for the aesthetic appreciation of nature and of art? In his book on the aesthetic appreciation of nature, the British philosopher Malcom Budd claims that the mistaken search for a model of the correct or appropriate aesthetic appreciation of nature reflects a lack of recognition of the freedom that is integral to the aesthetic appreciation of nature. His proclamation that the freedom of nature is an aspect of nature's distinctive aesthetic appeal[5] leads him to the conclusion that the aesthetic appreciation of nature is endowed with a freedom denied to the appreciation of art, since there are historically formed models for the correct or appropriate aesthetic appreciation of art.[6] The existence of such rules is undeniable. For example, it can be generally accepted that we have to sit still and watch a screen, if we want to understand a movie—that we can enjoy a ballet, without touching the dancers—but that we might try do jump on stage during a rock concert. In spite of that, some works of art deliberately stimulate us to experiment with the rules of the correct or appropriate appreciation. Artwork that can be understood as open processes, that breaks up the traditional distinction between spectator and artist. Therefore we might find that some art is defined by distinct rules of appreciation, while other artwork—instead of respecting models for the correct or appropriate aesthetic appreciation of art—challenges these

---

[4] Düwell (2000: 20).
[5] Budd (2002: 147–148).
[6] Budd (2002: 147–148).

established rules. Therefore, the rules that determine our approach on art are relative.

Likewise there are definite conditions for the aesthetic appreciation of nature. For example, if we want to listen to a particular bird sing—we have to locate him in his natural habitat, most likely very early in the morning when birds tend to sing—to enjoy the melody of his singing. Or to give another example, we can appreciate the roughness of the Atlantic Ocean—while gazing at it harbour-side, while studying an impressionist painting or maybe while enjoying a regatta on television—all within the same spectator role. Each of these scenarios encourages a different attitude, related to different modes of correct aesthetic appreciation. In addition, the rules for correct appreciation are based on a distinction between nature and art that is relative to a social context: The common elements found in the autonomic nature appreciated by a fisherman in Jakarta and the autonomic nature appreciated by a lawyer in Istanbul might be difficult to establish. The conceptual differences are certainly important and are partially dependent on the social relativity of correct modes of aesthetic appreciation. In spite of this, I propose universal patterns for the aesthetic appreciation of nature that are based on the same rules established for the aesthetic appreciation of art.

Imagination and autonomous nature

Can autonomous nature exist without human being? Autonomic nature is a product of our imagination. It is not a part of physical nature, but physical nature can give us an impression of autonomic nature. In fact, we can experience autonomic nature on the condition that we project it onto physical nature. Physical nature can give us the impression of independence, but this independence is nothing but a reflection of our own independence onto nature. Hence the independence of nature is created by human activity.

This seems to be contradictive. Why should the autonomy of nature depend on human activity that is just a detail of the general development of the planet earth? According to scientific research, life developed for millions of years before the first mammals appeared and in a way, the birth of the whole universe can be understood as a result of a natural development. We certainly claim to know that the existence of life is due to a natural development that took place long before human beings were present on earth. We might even argue, that human presence on earth challenges the independence of nature. Independence is a human value imposed on nature. The possibility to appreciate independent nature relies on a reflexive activity that is a condition for the definition of any concept. Our knowledge is conditioned by our ability to handle abstract symbols. This ability allows us to take a distance from nature, which is a condition for aesthetic appreciation. Without this distance, we would interact with our natural environment and therefore we would not be able to form a concept of autonomic nature. Any speculation about a past or future development of nature would be impossible. Therefore, the existence of autonomic nature results from a specifically human way of life that implies the ability to create art. It is created by our imagination that is inspired by art. As imagined autonomy is not effective autonomy, the importance of the aesthetic appreciation of nature for the definition of ethic rules becomes questionable. Is it possible to establish universal ethic patterns for the creation of our environment in reference to the relative standards of the aesthetic appreciation?

Ethical implications

Can autonomous nature guide our behaviour towards existing physical nature nevertheless? The aesthetic appreciation of autonomous nature has ethic significance. Indeed, the relationship between autonomous nature and our aesthetics preferences is a significant factor in forming our ethical values regarding both, nature and art. The importance of autonomous nature for our ethical values becomes clear when we take into consideration the interdependence between autonomous nature and art. Artistic creativity creates autonomic nature and the capacity to imagine autonomic nature is a condition for artistic creativity. Therefore, the ethical significance of art depends on our capacity to imagine autonomic nature.

Links between ethics and aesthetics

Are aesthetic values limited by their ethic significance? Events that provide aesthetic pleasure are not always morally right. Neither will they necessarily be helpful to form moral characters. Ethical and aesthetical approaches to nature seem to be opposed or at least deal with very different sets of issues. The philosopher Josef Früchtl points out that we can distinguish at least five basic models to understand the relationship between ethics and aesthetics.[7] Either there is a mutual influence either there is not. If we agree, that there is a link between aesthetics and ethics, we face some more choices. Either ethics can determine aesthetics or aesthetics can determine ethics or ethics and aesthetics can be equal

---

[7] Früchtl (1996: 59–77).

"partners" or they can complete each other in various aspects. The philosopher Martin Seel affirms this last possibility. Instead of following the Kantian idea of a link between natural beauty and morality, he shows that art can provide us with hints about our moral choices, even when these hints might not lead to moral imperatives.

The emotional connection between aesthetic and moral values

Is there an emotional connection between aesthetic and moral values? According to Kant the ethic value of art is weak. For him it is the interest in natural beauty that indicates the presence of a moral feeling. The pure judgement of taste is not about the matter but the form of an object. The moral judgement on the contrary is not about the matter but the form of a principle. The aesthetic judgement is linked to universal pleasure, while the moral judgement is linked to an ethical standard. Both kinds of judgements represent a pleasure but are also independent from a subjective will. This position includes an analogy between a moral feeling and a feeling for natural beauty. Kant asserts that the harmonious operation of sensibility and understanding in the experience of beauty promotes the mind's receptivity to moral feeling, making the concept of a purpose of nature a suitable mediating link between the realms of nature and freedom.[8] The Kantian link between beauty and moral is certainly weak. Kant admits this himself by explaining that the pleasure in beauty gives us freedom from any mere sensual enjoyment; but that this freedom is still represented more in play than in exercising a law-ordained function which is the genuine characteristic of human morality.[9] Budd, introduced earlier on, concludes that there are more effective ways of fostering the development of morality in someone than through encouraging him to love natural beauty. That may be so, but it doesn't contradict the emotional link between aesthetic and moral values. To the contrary, it seems to speak for an emotional connection rather than the rational one implied by the regulation of morality through law. In fact, our ethic and aesthetic evaluation of the environment is related to the sensibility of our body. The impact of nature and art on our sensibility can integrate both: immediate sensation and conceptual distance. Our body is the link between the environment and us, a sort of cargo turnaround for culture and art.

---

[8] Kant (1992: 86).

[9] *Ibid.* 194.

Aesthetic pleasure as a part of a good life

Is there an inter-subjective ethical value that resides in our appreciation of aesthetic nature? Seel proposes that the ethical implications of the aesthetic experience of nature diverge from an ethic of 'right' behaviour in Kantian terms.[10] His concern is based on setting an ethic of 'good life'[11] or more precisely the structure of satisfying 'successful existence' along the lines of Greek ethical thought. I will from now on use the concept 'ethical' to describe a structure that favours satisfying ways of living, which might nevertheless lead to 'right' norms for social behaviour.

Is a good life related to aesthetic pleasure? Seels argument against the unity of the aesthetical and the ethical approach, is based on a distinction between a 'rich' and a 'humble' concept of good life.[12] The rich concept of good life implies an aesthetic dimension of life whereas a humble concept will not. This does not mean that only the aesthetic approach towards life can be good. Still, aesthetic experience can inform and enrich good life. The aesthetic experience of nature is an encounter with a form of good life and for this reason it can provoke strong ethical experiences. If aesthetic nature helps us to live in a reality that leaves room for intensified sensation, diversified perception and distance from the limitations of our everyday life, it certainly has an importance for the notion of an ethically good life. Consequently, Seel distinguishes three contexts in which aesthetic experience intervenes with ethical human existence. First, it benefits our involvement with inter-subjective forms of good life. Second, it enables us to gain critical distance from our life. Lastly, it opens prospects to other possible ways of living. The link between aesthetic and ethical values allows me to finally offer concluding remarks about the ethical impact of aesthetic experience on a moral attitude towards nature.

Good life and universal norms for our behaviour

How does the link between aesthetic pleasure and a good life affect our moral behaviour toward nature? Even if the aesthetic autonomy of nature can be considered an irreversible fact, the physical nature onto witch we can project independent nature is not indestructible. According to Seel, 'technical art' has the capacity to either destroy or develop autonomy in physical nature. Technical art intervenes in the structural development of

---

[10] Seel (1991).
[11] *Ibid.* 289.
[12] *Ibid.* 331.

nature and therefore is characterized—at least to some extent—by the attempt to control nature. In opposition, the approach of aesthetic art towards nature depends—as I have pointed out—on the existence of an autonomous nature. Aesthetic nature enables us to imagine variations of reality. Seel puts it in the following way: imagination can create 'space within space' and 'time within time' when it's projected onto nature. Aesthetic nature stimulates reflection and therefore facilitates the search for a satisfying good life. This explains the ethical value of autonomous nature for individual everyday life. The existence of physical nature is a necessary condition for the possibility of autonomous nature, which is itself vital for the aesthetic experience of nature, therefore a 'rich' and 'good' life. The junction of the ethical and the aesthetical leads Seel to the moral proposition that the conservation of autonomous nature is a sign of respect for individual existence. It follows that the ethical value of autonomous nature derives from a personal interest. This personal interest relates to the moral demand for respect towards ways of life that are advantageous for every human being. If we want to satisfy the demand of individual freedom and satisfaction, the appreciation of autonomous nature inspires the defence of physical nature in our environment. Therefore aesthetic appreciation can be a helpful mechanism to inform and influence the ethical values that could benefit the protection of natural environments.

The specific interest of aesthetic pleasure for our moral behaviour towards nature

What is the specific interest of aesthetic pleasure for our moral behaviour towards nature? Reflection on autonomous nature implies evaluative and universal norms for human behaviour. What makes these norms specific is that they are not instrumental. Aesthetic appreciation can provide us with sensations that we would not have otherwise, if we were only to look upon nature as a social, a scientific, or a mythical object. Of course this does not mean that moral considerations concerning nature always depend on aesthetic experience. Aesthetic nature is not the only 'valuable' nature of value. Moral norms for human behaviour towards nature can also be founded on ecological, economical, medical, historical or other reflections. Still, aesthetic considerations are a useful addition to these approaches. They are particularly helpful in extending the limits of our imagination when feeling the emotional impact of autonomic nature in our life.

What does this mean for our behaviour towards nature? The autonomy of nature that inspires artistic creativity cannot be found in physical nature. However, the possibility to imagine autonomic nature depends on our

Royal Historical Society Bibliography and sits on the Editorial Committees of *Cultural and Social History* and *Twentieth Century British History*. His research on wartime Britain and representations of the Home Front has led him to take an increasing interest in the photography of Bill Brandt beyond his well-known 'Blitz reportage'.

CHRISTIAN W. DENKER Studied aesthetics and philosophy at the University of Hamburg. His doctoral thesis was on the link between life and the art of Jochen Gerz at the University of Paris 1 (Panthéon/ Sorbonne). Christian currently teaches at the Department of Aesthetics and Science of Art at the University of Paris 1 and at the Institute of Philosophy at the University of Vienna. Christian has published pieces on nature, art, aesthetics and digestion.

JULIENNE HANSON (Dip.Arch, M.Sc., Ph.D.) is Professor of House Form and Culture at the Bartlett, University College London (UCL), where she has been a teacher and researcher since 1975. She has recently profiled the UK's housing stock with the needs of older people in mind, investigated the layout and design of residential care homes in relation to quality of life issues and completed two linked studies of the housing, support and care needs of older people with visual impairments and the housing and support needs of vision impaired adults of working age. Current projects include mainstreaming 'telecare' services in the homes of older people, the design of culturally sensitive homes, remodelling sheltered housing and residential care homes to 'extra care' housing the provision of inclusive and sustainable public toilets in city centres.

THOMAS HEYD obtained his Ph.D. at the University of Western Ontario, Canada, and also studied in Germany and Spain. He teaches philosophy and environmental studies at the University of Victoria (Canada). His publications include *Recognizing the Autonomy of Nature* (2005, Columbia University Press), *Encountering Nature: Toward an Environmental Culture* (2007, Ashgate Publishing), *Environmental Consciousness and Action* (1996, co-author with Wolf Schluchter, German Environment Ministry), and *Aesthetics and Rock Art* (2005, co-editor with John Clegg, Ashgate).

EDWARD H. HUIJBENS is currently the director of the Icelandic Tourism Research Centre. He studied geography at the University of Iceland and later cultural geography at Durham University in England, completing an M.A. in 2001 and Ph.D. in 2006. He is the author of articles in numerous scholarly journals in both Iceland and internationally and has previously co-edited *Technology in Society/Society in Technology* (2005, University of Iceland Press).

Paola Jiron is a Chilean scholar from the Faculty of Architecture and Urbanism at the University of Chile where she has carried out work mainly in the areas of housing policy, urban quality of life, residential satisfaction, and gender in human settlements. She holds a B.Com. from Concordia University, Canada and an M.Sc. from University College London, UK. She is currently finishing her Ph.D. in Urban and Regional Planning at the London School of Economics and Political Science.

Shauna Laurel Jones lives in Reykjavík and writes on contemporary Icelandic art, film and music. Her publications include articles for The Centre for Icelandic Art's *LIST* magazine as well as *The Reykjavík Grapevine,* and she has written entries on 20th-century Icelandic visual artists and composers for the forthcoming *Encyclopedia of Contemporary Scandinavian Culture*. She studied art history at Oberlin College and the University of California, Santa Barbara.

Ólafur Páll Jónsson is Lecturer in Philosophy at the Iceland University of Education. He studied philosophy at the University of Calgary and at MIT before returning to Iceland. Ólafur Páll is an editor of *Art, Ethics and Environment: A Free Inguiry into the Vulgarly Received Notion of Nature* (2006, Cambridge Scholars Press) and *Náttúra vald og verðmæti* ('Nature, Power and Values' 2007, Icelandic Literary Society). He is also a member of the board of the Icelandic Nature Conservation Association.

Doreen Massey is Professor of Geography at the Open University, UK. She writes on space and place, their conceptualisation and its importance for both theory and politics, and on globalisation, cities and regions. Her most recent books are *For Space* (2005, Sage) and *World City* (2007, Polity).

Julia Nevárez obtained her Ph.D. in Environmental Psychology at the Graduate Centre of the City University of New York and lives in Harlem, New York City. She is currently Assistant Professor in the Sociology and Anthropology Department at Kean University, New Jersey. Her research focuses on urban issues, globalization, urban development, public space and techno-culture most recently.

KENICHI NONAKA did his undergraduate and M.A. studies in geography at Nagoya University, Japan. He later completed a Doctor of Science in Kyoto University, Japan. He is currently a Professor of Geography at Rikkyo University.

GÍSLI PÁLSSON is a Professor of Anthropology at the University of Iceland. He studied anthropology at the University of Iceland and later at the University of Manchester, completing his doctorate in 1982. Returning directly to the faculty at the University of Iceland he became a professor in 1992. With recent publications such as *Anthropology and the New Genetics* (2007, Cambridge University Press) and *Travelling Passions – the Hidden Life of Vilhjálmur Stefánsson* (2006, University of Manitoba Press), Gísli's research focuses on the social implications of biotechnology and concerns about the collection, storing, and exchange of human bodily material and medical information. Also, he is engaged in research on arctic cultures, human-environmental relations, and Inuit genetic history. Gísli has done anthropological fieldwork in Iceland and The Republic of Cape Verde.

ELISEBHA FABIENNE PLATZER is currently a Ph.D. candidate in Communication Sciences, University of Turin (Italy), in cooperation with international Ph.D. Program Leopold-Franzens-Universität Innsbruck Department of Romance Languages, Innsbruck (Austria). Her Ph.D. thesis is titled *Italo Calvino e Le citta`invisibili. Visioni e percorsi artistici* and focuses on the semiotic literary structures in Calvino's novel and their intersemiotic translation into arts and architecture. In 2006 she was Research Fellow at Brown University, Department of Italian Studies, Providence (Rhode Island). In 2004 she received her M.A. with honours in Italian and French Studies with teacher training program, Leopold-Franzens-Universität Innsbruck. She conducted research in different Archives and Libraries: Harvard Widener Library, Archivio di Stato (Turin), Archivio Einaudi (Turin), Sorbonne Nouvelle III (Paris), Universita`degli Studi di Venezia Ca' Foscari (Venice). She studied for one year in 2000 at the Universita`delgi Studi di Napoli Federico II (Naples).

ANNA LÍSA RÚNARSDÓTTIR studied architecture at the Ecole d'Architecture de Grenoble in France before completing a Graduate Diploma in World Art Studies at the University of East Anglia in the UK. In her doctoral work she combined her background in architecture with a focus on material culture studies within anthropology. She carried out research in the Kingdom of Tonga on the relationship between social change, cultural values and recent transformations of houses and material culture of the home. She was awarded her Ph.D. from the Department of Anthropology at University College London in 2004. Dr. Rúnarsdóttir is currently Head of Research and Collections at the National Museum of Iceland, where she has carried out research on Icelandic vernacular architecture.

HRAFNHILDUR SCHRAM has a filosofie licentiante degree in history of art from the University of Lund, Sweden. She is the former director of the National Museum of Einar Jónsson in Reykjavík as well as the National Museum of Ásgrímur Jónsson. She has written books and made documentary films on Icelandic art and pictorial artists. She lives in Reykjavík and is currently working as a writer.

MICHAEL SHIELL is an Australian based arts researcher at the University of Ballarat. Along with his own artistic creative endeavours he is presently working on his Ph.D. exploring alternative means of documenting ephemeral environmental art.

ANDREW A. STONYER is Professor in Fine Art at the University of Gloucestershire, UK. He is a practicing artist whose work is concerned with kinetics and interactive patterns of movement. He studied at Northampton School of Art, Loughborough College of Art and Design, the Architectural Association, Leicester Polytechnic and the Slade School of Fine Art. In 1978 he was awarded the first practice based Ph.D. in Fine Art in the UK. He has taught in Holland, Canada, USA and Turkey, where he has also exhibited and completed commissions.

MARIA TAMBOUKOU is Reader in Sociology and Co-director of the Centre of Narrative Research in the Social Sciences at the University of East London. After obtaining a B.A. in Law at the University of Athens, she went on to study for an M.A. and a Ph.D. in the sociology of education at King's College London. Her research activity is in the areas of Foucauldian and Deleuzian analytics, critical feminisms and auto/biographical narratives. She is the author of *Women, Education and the Self: a Foucauldian Perspective* (2003, Palgrave, Macmillan) and co-editor with Stephen J. Ball of *Dangerous encounters: genealogy and ethnography* (2004, Peter Lang).

SERGE TAMPALINI (Ph.D.) currently lectures in Theatre Studies and Design at Murdoch University, Western Australia, where he specialises in performance theory, aesthetics and theatre semiology, and actor training. Serge's theatre work has been seen in all Australian capital cities as well as Morocco, Malta and Canada.

JESSICA THOM uses photography to make images that explore how our environment is seen and experienced. Based in the UK Jessica also works as an artist educator for galleries, schools and community groups, specialising in working with children and young people with disabilities. Educated at Camberwell College of Art, The Princes Foundation and the Royal College of Art, Jessica's current practice involves working in collaboration with a space's users to create accessible articulate portraits of our surroundings. Jessica has exhibited all over the UK including at the Hoopers Gallery in London, the Chapel Gallery in Lancashire and at URBIS in Manchester in 2005 she was awarded the Hooper's Award.

GIOVANNI TUZET studied law and philosophy in Turin and Paris and wrote his Ph.D. thesis on Peirce's theory of inference. Formerly post-doc researcher at the universities of Lausanne, Switzerland, and Ferrara, Italy, he presently teaches Philosophy of Law at Bocconi University in Milan, Italy. His areas of interest include epistemology, philosophy of logic, normative sciences and, in particular, philosophy of law.

NOZOMI YANAHARA completed a Bachelor of Economics degree from Nagoya University, Japan, but currently works as a manga artist.

REEM ZAKO (B.Sc., M.Sc.) is trained as an architect and is a researcher at the Bartlett School of Graduate Studies/UCL. She has been involved in a number of projects exploring the relation between the built environment, at the levels of home, neighbourhood and the urban realm with the quality of life and inclusion of its residents. Her current research agenda is focused on the issue of housing and social inclusion. Her broader research interests lie in the social constructions of the built environment, and the social dimensions of the spatial configurations of the neighbourhood and the home.